49 Topics in Current Chemistry
Fortschritte der chemischen Forschung

Symmetry and Chirality

Springer-Verlag
Berlin Heidelberg New York 1974

This series presents critical reviews of the present position and future trends in modern chemical research. It is addressed to all research and industrial chemists who wish to keep abreast of advances in their subject.

As a rule, contributions are specially commissioned. The editors and publishers will, however, always be pleased to receive suggestions and supplementary information. Papers are accepted for "Topics in Current Chemistry" in either German or English.

Any volume of the series may be purchased separately.

ISBN 3-540-06705-1 Springer-Verlag Berlin Heidelberg New York
ISBN 0-387-06705-1 Springer-Verlag New York Heidelberg Berlin

This work is subject to copyright. All rights are reserved, whether the whole or part of the material is concerned, specifically those of translation, reprinting, re-use of illustrations, broadcasting, reproduction by photocopying machine or similar means, and storage in data banks. Under § 54 of the German Copyright Law where copies are made for other than private use, a fee is payable to the publisher, the amount of the fee to be determined by agreement with the publisher. © by Springer-Verlag Berlin Heidelberg 1974. Library of Congress Catalog Card Number 51-5497. Printed in Germany. Typesetting and printing: Hans Meister KG, Kassel. Bookbinding: Konrad Triltsch, Graphischer Betrieb, Würzburg.

The use of registered names, trademarks, etc. in this publication does not imply, even in the absence of a specific statement, that such names are exempt from the relevant protective laws and regulations and therefore free for general use.

Contents

Permutation Groups, Symmetry and Chirality in Molecules
C. A. Mead 1

Editorial Board:

Prof. Dr. A. Davison	Department of Chemistry, Massachusetts Institute of Technology, Cambridge, MA 02139, USA
Prof. Dr. M. J. S. Dewar	Department of Chemistry, The University of Texas Austin, TX 78712, USA
Prof. Dr. K. Hafner	Institut für Organische Chemie der TH D-6100 Darmstadt, Schloßgartenstraße 2
Prof. Dr. E. Heilbronner	Physikalisch-Chemisches Institut der Universität CH-4000 Basel, Klingelbergstraße 80
Prof. Dr. U. Hofmann	Institut für Anorganische Chemie der Universität D-6900 Heidelberg 1, Im Neuenheimer Feld 7
Prof. Dr. J. M. Lehn	Institut de Chimie, Université de Strasbourg, 1, rue Blaise Pascal, B. P. 296/R8, F-67008 Strasbourg-Cedex
Prof. Dr. K. Niedenzu	University of Kentucky, College of Arts and Sciences Department of Chemistry, Lexington, KY 40506, USA
Prof. Dr. Kl. Schäfer	Institut für Physikalische Chemie der Universität D-6900 Heidelberg 1, Im Neuenheimer Feld 7
Prof. Dr. G. Wittig	Institut für Organische Chemie der Universität D-6900 Heidelberg 1, Im Neuenheimer Feld 7

Managing Editor:

Dipl.-Chem. F. Boschke	Springer-Verlag, D-6900 Heidelberg 1, Postfach 1780
Springer-Verlag	D-6900 Heidelberg 1 · Postfach 1780 Telephone (06221) 49101 · Telex 04-61723
	D-1000 Berlin 33 · Heidelberger Platz 3 Telephone (030) 822011 · Telex 01-83319
Springer-Verlag New York Inc.	New York, NY 10010 · 175, Fifth Avenue Telephone 673-2660

Permutation Groups Symmetry and Chirality in Molecules

Prof. C. Alden Mead

Chemistry Department, University of Minnesota, Minneapolis, Minnesota, USA

Contents

I.	Introduction	3
II.	Mathematical Background	6
	A. Assumed Background; Notation	7
	B. Group Algebra and Projection Operators	8
	1. Group Algebra	8
	2. Regular Representation	9
	3. Projection Operators	9
	C. Induced and Subduced Representations	15
	1. Subduced Representations	15
	2. Induced Representations	15
	D. Symmetric and Hyperoctahedral Groups	19
	1. Definitions and General Remarks	19
	2. Classes	21
	3. Representation Theory	24
	E. Regular Induction from $\mathfrak{S}_n$ and $\mathfrak{S}_n^!$ to $\bar{\mathfrak{S}}_n$	33
	F. The Transfer Condition	36
III.	Chirality Functions; Qualitative Completeness	43
IV.	Simple Explicit Forms of Chirality Functions	51
V.	Active and Inactive Ligand Partitions	59
VI.	Class-Specific and Ligand-Specific Chirality Functions	62
VII.	Chirality Numbers	64
	A. Achiral Ligands	64
	B. Chiral Ligands	65

1

VIII.	Homochirality	68
IX.	Some Experimental and Theoretical Applications	72
X.	Unsolved Problems; Discussion	76
XI.	Appendix: Results for Some Special Skeletons	78
	A. Skeletons Considered	78
	B. Chirality Functions	79
	C. Chiral Representations for 6-site Skeletons	83
	D. Properties of Chiral Representations	84
	E. Properties of Skeletons	85
XII.	References	86

I. Introduction

It has long been known that a molecule which is not superimposable with its mirror image is capable of rotating the plane of polarized light, with mirror image molecules rotating the plane through the same angle in opposite directions. This optical rotation is one example of a "chiral" or "pseudoscalar" molecular property, *i.e.*, of a numerical property which is unaffected by changes in orientation of the molecule but which reverses its sign when the molecule is replaced by its mirror image. Molecules possessing such properties, which are therefore different from their mirror images, are called chiral; molecules superimposable with their mirror images are achiral.

A molecule can always be pictured as a skeleton or frame providing sites to which ligands have been attached. If the skeleton itself is achiral, then chiral properties of molecules arise from differences between the ligands. It is natural, therefore, to try to account for such properties, either theoretically or empirically, in terms of differences in various properties of the ligands. For example, in the case of the methane skeleton consisting of the carbon atom with its four tetrahedrally directed bonds, it is well known that a molecule is chiral only if all four ligands are different. Reasoning from this circumstance, Crum Brown [1] and Guye [2] proposed as long ago as 1890 that the optical rotatory power might be proportional to a "product of asymmetry" of the form

$$\Theta = (a-b)(b-c)(c-d)(a-c)(a-d)(b-d). \tag{1}$$

If any two of the quantities a, b, c, d in Eq. (1) are equal, then $\Theta = 0$. If the values of any two of them are interchanged, Θ changes sign. If a, b, c, d are identified with some property of the ligands, therefore, then Θ is a chiral property of the molecule, and is thus a candidate for an approximate representation of the rotatory power. Crum Brown and Guye identified the quantities a, b, c, d with the *masses* of the ligands.

In 1934, Boys [3] proposed a molecular model which, as he was able to show, possessed optical activity. His formula for the rotatory power was considerably more complicated than (1), but contained a factor of the same form as (1). In his case, however, the quantities a, b, c, d were identified with the *radii* of the ligands. One could cite other examples of models leading to formulae containing factors of this form.

The function Θ in (1) may or may not accurately describe an experimentally interesting chiral property; it does, however, possess the necessary *symmetry* properties for doing so, and therefore is an example of what is called a "chirality function" (to be defined precisely in Section III). Its form was arrived at basically through symmetry considerations, but not in a way that would systematically yield analogous formulas for other skeletons. The question arises of just how much can be said about chirality functions in a systematic way based on symmetry considerations alone. It is to be expected that group theory should play a major role in any study of this question.

The systematic group-theoretical study of chirality functions was taken up in 1967 by Ruch, Schönhofer, and Ugi [4], and further pursued by Ruch and Schönhofer [5] in 1968. While of considerable importance at the time, these efforts have now been largely superceded by the elegant, definitive form given the theory in 1970 by Ruch and Schönhofer [6]. In this theory, a crucial role is played by the transformation properties of the chirality functions, not just under the point group of the skeleton, but under the larger group of all permutations of the ligands among the n sites of a skeleton (symmetric group). This group generates not only all orientations of a molecule and its mirror image which leave the skeleton invariant, but all isomers as well. More recently, in a piece of work in which the present author also participated, the theory has been further generalized to permit the ligands themselves to be chiral.[7] In this case, the group which generates all the isomers is the "hyperoctahedral" group consisting of all combinations of permutations of ligands among the sites and conversion of individual ligands into their mirror images.

The assimilation of this work on the part of the chemical community has been hindered somewhat by the fact that the theory makes use of aspects of group theory which are not part of the stock-in-trade of the chemist, even of the chemist who regularly uses group theory for standard purposes such as classification of energy levels and of normal modes, selection rules, etc. In particular, the comprehension of this theory requires a knowledge of the representation theory of the symmetric and hyperoctahedral groups, and of the concept of induced representations and their properties. While excellent expositions of these topics are to be found in the mathematical literature, they are usually formulated in a manner foreign to the chemist's version of group theory, and thus relatively inaccessible to him.

The present article is an attempt to bridge this mathematical gap, and also to present the theory as far as possible in a unified form, including together the cases of achiral and of chiral ligands. It is hoped that it will help to make the theory more easily accessible to readers with only a normal chemist's knowledge of group theory. Up to now, the only exposition of this work addressed to such readers has been the short account by Ruch [8]. While an excellent informal introduction to some of the ideas, however, Ruch's article makes no pretense of developing the full theory in a systematic way. Thus, it is believed that the present article can serve a useful purpose.

In Section II, the longest single section in the paper, the necessary group-theoretical background is developed, starting with group-theoretical concepts and results which are reasonably well-known to chemists, or at least readily accessible to them. For the reader who is interested in results but not in proofs, it is indicated at the beginning of Section II which parts of the section are essential for an understanding of the sequel and which may be skipped. The subsequent sections develop the theory up to its present stage, and also include discussion of some applications which have been made. An appendix contains a compilation of results which have been obtained for a number of specific skeletons.

The formulation, while equivalent to that of the original articles, is not in all places identical. Some of the changes made are intended to be in the interests of simplicity or clarity, while others have the intent of bringing the notation into conformity with that used elsewhere in the literature.

Equations are numbered anew in each (Roman numeraled) section. Within each section, equations from within that section are referred to simply by their numbers, while equations from other sections are designated by the equation number preceeded by a Roman numeral for the section. Thus, Eq. (3) of Section II would be referred to as (3) within Section II, but in other Sections as (II—3).

II. Mathematical Background

In this section, the group theoretical background needed for the comprehension of the main body of the article will be developed. All the results derived here are also to be found in the mathematical literature [9,10]; however, the formulations usually used are somewhat forbidding to the chemist, being couched in language different from that used in chemical group theory. We will assume that the reader is familiar with the standard results of the chemist's group theory, as found, for example, in the books by Hamermesh [11], and Wigner [12], and will build from there.

In Subsection A we review without proof some of the standard results which we shall use; this will also serve to fix some of the notation. Subsection B deals with the group algebra, the regular representation, and projection operators, C with induction and subduction of representations, and D with the representation theory of the groups most important to us, the symmetric and hyperoctahedral groups. This is done in the Young diagram formulation, which is the one most suitable for our purposes. Subsection E deals with induction of representations between these groups, and F with certain relations between Young diagrams which we call the "transfer condition".

For the reader who is not interested in working through the proofs, but is mainly interested in results, the following is a guide, subsection by subsection, to those results most important for an understanding of the later sections.

A. Can be omitted, or just skimmed for definitions and notation.

B. Ditto.

C. Definitions of subduction, induction, and regular induction are important, as is the main theorem on induction embodied in Eq. (34), but not its proof.

D. Definitions of Young diagrams, tableaux, and operators should be understood, as well as the property of the Young operator of projecting onto an irreducible representation (Theorems 1 and 2).

E. The details of the induction prescription need not be mastered if the reader is willing to take some later results on faith; it is important to understand what the prescription accomplishes.

F. The meaning of the T-condition should be understood.

A. Assumed Background; Notation

A group will be denoted by a capital German letter ($\mathfrak{G}$), its order by the corresponding lower-case Latin letter (g), and its elements by lower-case Latin letters s, t, etc. The reader is assumed familiar with the concepts of subgroup and class, and with their elementary properties, *e.g.*, that the order of a subgroup, or of a class, must be a divisor of g. A class will be denoted by a script c, its order by c, sometimes with a subscript on both.

Familiarity is also assumed with the concepts of representation and irreducible representation (IR). A representation Γ of dimension n associates to each group element s an $n \times n$ matrix $D(s)$, with matrix elements $D(s)_{ij}$, in such a way that for every s, t, $D(s)D(t) = D(st)$, with the product formed by ordinary matrix multiplication. We will sometimes use the bra-ket notation

$$<i\,|D(s)|\,j>$$

for the matrix elements. Each representation can be decomposed into irreducible parts. The number of inequivalent irreducible representations is equal to the number of classes in the group, the sum of the squares of their dimensions equals the order of the group.

We shall make frequent use of the orthogonality relations:
If $\Gamma^{(\mu)}$, $\Gamma^{(\nu)}$ are two irreducible representations, then

$$\sum_{s \in \mathfrak{G}} D^{(\mu)}(s)_{ij}\, D^{(\nu)}(s^{-1})_{kl} = \frac{g}{n_\mu}\, \delta_{\mu\nu}\, \delta_{jk}\, \delta_{il}\,. \tag{1}$$

If the representations are unitary, (1) becomes

$$\sum_{s \in \mathfrak{G}} D^{(\mu)}(s)_{ij}\, D^{(\nu)}(s)^{*}_{lk} = \frac{g}{n_\mu}\, \delta_{\mu\nu}\, \delta_{jk}\, \delta_{il}\,. \tag{2}$$

The character of an element in a particular representation is defined by

$$\chi^{(\mu)}(s) = Tr\, D^{(\mu)}(s)\,.$$

It is obvious that the character is the same for all elements in a class, and that it is invariant under similarity transformations of the representation. For the characters, the orthogonality relations (1) and (2) take the form

$$\sum_s \chi^{(\mu)*}(s)\, \chi^{(\nu)}(s) = g\, \delta_{\mu\nu}, \tag{3}$$

or

$$\sum_c c\, \chi^{(\mu)*}(c)\, \chi^{(\nu)}(c) = g\, \delta_{\mu\nu}. \tag{4}$$

An orthogonality relation reciprocal to (4) also holds, namely

$$\sum_\mu \chi^{(\mu)*}(c)\, \chi^{(\mu)}(c') = \frac{g}{c}\, \delta_{cc'}, \tag{5}$$

where the sum goes over all inequivalent irreducible representations.

This completes the summary of the assumed background. We now proceed to develop the special tools needed in this paper which are not part of the chemist's standard group theoretical repertoire.

B. Group Algebra and Projection Operators

1. Group Algebra

It is customary in mathematical treatments of group theory to develop the representation theory entirely in terms of the group algebra.[9] Our procedure will be to use those aspects of representation theory which we already know by other means to help in developing the theory of the algebra.

Given a group $\mathfrak{G}$, the group algebra $\mathfrak{U}$ is defined as consisting of all possible symbolic sums of group elements multiplied by complex numbers. Thus, an element a of the algebra is written

$$a = \sum_s a(s)\, s, \tag{6}$$

where the $a(s)$ are complex numbers. Addition and multiplication of elements of the algebra are defined as one would expect: for two elements a and $b = \sum_s b(s)\, s$, the sum is defined as

$$a + b = \sum_s [a(s) + b(s)]\, s \tag{7}$$

and the product (in general not commutative) by

$$ab = \sum_{s,t} a(s)\, b(t)\, st$$

$$= \sum_s \left[\sum_t a(t)\, b(t^{-1}s)\right] s. \tag{8}$$

Given a representation Γ of $\mathfrak{G}$, associating s with $D(s)$, we can form a representation of the algebra by associating to each element a the matrix $D(a) = \sum_s a(s) D(s)$. It is evident that addition and multiplication of the matrices so defined will reproduce the rules (7) and (8).

2. Regular Representation

A particular representation which we shall need is the "regular" representation of dimension g, which symbolically uses the group elements themselves as basis vectors of a "permutation representation".

Since multiplication by the element s transforms each element of into another element, with no two going into the same element, it can be thought of as effecting a permutation of the group elements. This can be represented in matrix form by defining

$$D^{(R)}(s)_{tu} = 1 \text{ if } su = t$$
$$= 0 \text{ otherwise.} \qquad (9)$$

The representation so defined is the regular representation $\Gamma^{(R)}$. It has dimension g, and each row and each column of any $D^{(R)}(s)$ has exactly one element "1", with the rest being zero. Only the unit element, 1, has diagonal matrix elements. Thus, $\chi^{(R)}(1) = g$, with the other characters being zero.

3. Projection Operators

The irreducible representations of $\mathfrak{G}$ may be used to define a set of elements of the group algebra called "projection operators". The projection operator $e_{ij}^{(\mu)}$ associated with the μ'th irreducible representation is defined by

$$e_{ij}^{(\mu)} = \frac{n_\mu}{g} \sum_s D^{(\mu)}(s^{-1})_{ji} s . \qquad (10)$$

For each irreducible representation $\Gamma^{(\mu)}$ there are n_μ^2 different projection operators, so in all there are $\sum n_\mu^2 = g$ such operators. As we shall see presently, they are linearly independent, and thus form an alternative basis for the group algebra: any element of $\mathfrak{U}$ can be represented as a linear combination of the $e_{ij}^{(\mu)}$.

The product of two projection operators can be worked out as follows:

$$e^{(\mu)}_{ij} e^{(\nu)}_{kl} = \frac{n_\mu n_\nu}{g^2} \sum_{st} D^{(\mu)}(s^{-1})_{ji} D^{(\nu)}(t^{-1})_{lk} \, st \, . \tag{11}$$

Now, if we define $st = r$, $t = s^{-1}r$, $t^{-1} = r^{-1}s$, (11) becomes

$$e^{(\mu)}_{ij} e^{(\nu)}_{kl} = \frac{n_\mu n_\nu}{g^2} \sum_r \{ \sum_s D^{(\mu)}(s^{-1})_{ji} D^{(\nu)}(r^{-1}s)_{kl} \} \, r \, . \tag{12}$$

Because of the representation property of the D matrices we have

$$D^{(\nu)}(r^{-1}s)_{lk} = \sum_m D^{(\nu)}(r^{-1})_{lm} D^{(\nu)}(s)_{mk}$$

which, when substituted into (12), gives

$$e^{(\mu)}_{ij} e^{(\nu)}_{kl} = \frac{n_\mu n_\nu}{g^2} \sum_r \{ \sum_{s,m} D^{(\mu)}(s^{-1})_{ji} D^{(\nu)}(s)_{mk} D^{(\nu)}(r^{-1})_{lm} \} r \, . \tag{13}$$

Summing over s and using the orthogonality relation (1), we find

$$e^{(\mu)}_{ij} e^{(\nu)}_{kl} = \frac{n_\mu}{g} \delta_{\mu\nu} \delta_{jk} \sum_r D^{(\mu)}(r^{-1})_{li} \, r = e^{(\mu)}_{il} \delta_{\mu\nu} \delta_{jk} \, . \tag{14}$$

To prove the linear independence of the $e^{(\mu)}_{ij}$, we set a linear combination of them equal to zero:

$$a^{(\mu)}_{ij} e^{(\nu)}_{ij} = 0 \, .$$

Multiply on the right by one of the $e^{(\mu)}_{ii}$, and on the left with $e^{(\mu)}_{jj}$, and use (14). This gives

$$a^{(\mu)}_{ij} e^{(\mu)}_{ij} = 0 \, .$$

Since the projection operators themselves are obviously not zero, it follows that all the coefficients $a^{(\mu)}_{ij}$ are zero, so the $e^{(\mu)}_{ij}$ are indeed linearly independent. It follows that they can be used as a basis for the group algebra. In particular, any group element s may be represented as

$$s = s^{(\mu)}_{ij} e^{(\mu)}_{ij} \, .$$

Multiplying by $e_{ii}^{(\mu)} \cdots e_{jj}^{(\mu)}$ and again using (14), we find[a]

$$s_{ij}^{(\mu)} e_{ij}^{(\mu)} = e_{ii}^{(\mu)} s e_{jj}^{(\mu)} . \tag{15}$$

If we substitute the definitions of the e operators into (15), it takes the form

$$s_{ij}^{(\mu)} e_{ij}^{(\mu)} = \frac{n_\mu^2}{g^2} \sum_{t,r} D^{(\mu)}(t^{-1})_{ii} \, tsr \, D^{(\mu)}(r^{-1})_{jj} . \tag{16}$$

By defining $u = tsr$, so that $r^{-1} = u^{-1}ts$, we can transform (16) into

$$s_{ij}^{(\mu)} e_{ij}^{(\mu)} = \frac{n_\mu^2}{g^2} \sum_{t,u} D^{(\mu)}(t^{-1})_{ii} \, D^{(\mu)}(u^{-1}st)_{jj} u . \tag{17}$$

Because the D matrices form a representation, we have

$$D^{(\mu)}(u^{-1}st)_{jj} = \sum_{k,l} D^{(\mu)}(u^{-1})_{jk} D^{(\mu)}(t)_{kl} D^{(\mu)}(s)_{lj} .$$

Substitution of this into (17), summation over t, followed by k and l, leads, with the help of the orthogonality relations, the representation property of the D, and the definition (10), to the result

$$s_{ij}^{(\mu)} = D^{(\mu)}(s)_{ij} . \tag{18}$$

As remarked above (cf. Eq. (9)), a representation of the group leads to a representation of the algebra; in particular, each of the projection operators will be associated with a matrix. If the representation is the irreducible one $\Gamma^{(\nu)}$, we find, using (9) and (2),

$$D^{(\nu)}(e_{ij}^{(\mu)})_{kl} = \frac{n_\mu}{g} \sum_s D^{(\mu)}(s^{-1})_{ji} D^{(\nu)}(s)_{kl}$$

$$= \delta_{ik} \, \delta_{lj} \, \delta_{\mu\nu} . \tag{19}$$

Thus, the matrix for $e_{ij}^{(\mu)}$ has only a single non-zero element, the ij'th, which is equal to one. The matrix for $e_{ij}^{(\mu)}$ vanishes for all irreducible representations except the μ'th.

---

[a] Multiplying by $x \cdots y$ means: multiply on the left by x and on the right by y.

Now, suppose that we have a representation decomposed into its irreducible parts, with basis vectors (in an obvious bra-ket notation) $|j^{(\mu)}>$. From the above considerations, we conclude that

$$e_{ij}^{(\mu)} |j^{(\mu)}> = |i^{(\mu)}>,$$

with all other basis vectors being annihilated by the projection operator. In particular,

$$e_{ii}^{(\mu)} |i^{(\mu)}> = |i^{(\mu)}>,$$

which justifies our use of the term "projection operators". In bra-ket notation, the e-operators may be written as

$$e_{ij}^{(\mu)} = |i^{(\mu)}><j^{(\mu)}| \; ; \; e_{ii}^{(\mu)} = |i^{(\mu)}><i^{(\mu)}|. \qquad (20)$$

The diagonal e-operators, $e_{ii}^{(\mu)}$, are special cases of what are called "primitive idempotents", which we now proceed to define. An operator $p \in \mathfrak{U}$ is called "idempotent" if it satisfies

$$p^2 = p. \qquad (21)$$

If we imagine an idempotent p expressed in terms of the e-operators, we see immediately that (21) must be satisfied separately for the parts belonging to each irreducible representation; hence, we can treat the parts independently in deducing consequences from (21). Consider, then, an idempotent $p^{(\mu)}$, belonging to the representation $\Gamma^{(\mu)}$. We define a "p-adapted basis" for $\Gamma^{(\mu)}$ as follows:

$$|j^{(\mu)}> = p^{(\mu)}|j_0^{(\mu)}>, \quad j = 1, 2, \ldots, h;$$

$$<j^{(\mu)}|p^{(\mu)}|m> = 0, \quad j = h+1, h+2, \ldots, n . \text{ all } m. \qquad (22)$$

In words, the first h basis vectors are obtainable by applying $p^{(\mu)}$ to other vectors, while the other $(n-h)$ are orthogonal to all of these. $h=0$ correspond to $p^{(\mu)} = 0$. In this basis, we see from (21), (22), that

$$p^{(\mu)} |j^{(\mu)}> = |j^{(\mu)}>, \quad j \leqslant h;$$

$$p^{(\mu)} |k^{(\mu)}> = \sum_{j=1}^{h} |j^{(\mu)}> p_{jk}, \quad k > h,$$

or

$$\begin{aligned}p^{(\mu)} &= \sum_{j=1}^{h} |j^{(\mu)}><j^{(\mu)}| + \sum_{j=1}^{h} \sum_{k=h+1}^{n_\mu} |j^{(\mu)}> p_{jk} <k^{(\mu)}| \\ &= \sum_{j=1}^{h} e_{jj}^{(\mu)} + \sum_{j=1}^{h} \sum_{k=h+1}^{n_\mu} p_{jk}\, e_{jk}^{(\mu)}.\end{aligned} \quad (23)$$

A general idempotent is a sum of operators of the form (23):

$$p = \sum_\mu p^{(\mu)}. \quad (24)$$

A *primitive* idempotent is now defined as an idempotent belonging to a single irreducible representation, and with $h = 1$. A primitive idempotent p, in the p-adapted basis, has the form

$$p = e_{11}^{(\mu)} + \sum_{k=2}^{n} e_{1k}^{(\mu)} p_k. \quad (25)$$

Using (14) and (25), we find, for primitive idempotent p,

$$p e_{st}^{(\mu)} p = \delta_{\mu\nu}\, \delta_{t1}\, (\delta_{s1} + p_s)\, p.$$

Since any $x \in \mathfrak{U}$ can be expressed in terms of the e's, it follows that, for primitive idempotent p and arbitrary $x \in \mathfrak{U}$,

$$pxp = f(x)p, \quad f(x) = \text{a number}, \quad f(1) = 1. \quad (26)$$

We now consider the converse of (26). Suppose we have a non-zero operator $q \in \mathfrak{U}$, with the property

$$qxq = f(x)q \quad \text{for all } x \in \mathfrak{U}, \text{ and}$$

$$f(1) \neq 0.$$

It follows immediately that $\bar{q} = \dfrac{1}{f(1)} q$ is an idempotent, and, according to (23), (24), can be expressed as

$$\bar{q} = \sum_\mu \{ \sum_{j=1}^{h_\mu} e_{jj}^{(\mu)} + \sum_{j=1}^{h_\mu} \sum_{k=h_\mu+1}^{n_\mu} e_{jk}^{(\mu)} p_{jk}^{(\mu)} \}. \tag{27}$$

Since $\bar{q} \neq 0$, we must have $h_\mu > 0$ for at least one μ. For such a μ, multiply $e_{11}^{(\mu)}$ by $\bar{q}\ldots\bar{q}$. We find, using (14), (27):

$$\bar{q} e_{11}^{(\mu)} \bar{q} = e_{11}^{(\mu)} + \sum_{k=h_\mu+1}^{n_\mu} e_{1k}^{(\mu)} p_{1k}^{(\mu)}. \tag{28}$$

The right-hand side of (28) is a primitive idempotent (cf. (25)), and can be a multiple of $\bar{q}$ only if $\bar{q}$ is itself a primitive idempotent. We have thus proved

Theorem 1: If a non-zero element q of $\mathfrak{U}$ has the property that $qxq = f(x)q$ for all x in $\mathfrak{U}$, and $f(1) \neq 0$, then q is a number times a primitive idempotent, and conversely.

Now suppose we have an operator $q \neq 0$ in $\mathfrak{U}$ with the property $q^2 = fq$. We wish to find an expression for f in terms of the coefficients in the expansion of $q = \sum q(s)s$. We do this by calculating $TrD^{(R)}(q)$ in the regular representation in two different ways:

(i) In the usual regular representation. Only $D^{(R)}(1)$ has diagonal elements, so $TrD^{(R)}(q) = gq(1)$.

(ii) Choose a q-adapted basis set in the space of $\Gamma^{(r)}$. Let h again be the number of vectors expressible as $q|m\rangle$. For $j \leq h$, we have $q|j\rangle = f|j\rangle$; for $k > h$, $\langle k|q|k\rangle = 0$. In this basis, therefore, $TrD^{(R)}(q) = fh$.

Equating the two values of the trace, we see that

$$f = \frac{g}{h} q(1),$$

which is the desired result. This, together with Theorem 1, leads immediately to

Theorem 2: If an element q of $\mathfrak{U}$ with $q(1) \neq 0$ has the property that $qxq = f(x)q$ for all $x \in \mathfrak{U}$, then q is a number times a primitive idempotent, and conversely.

Finally we establish a criterion for deciding whether two e-operators belong to the same irreducible representation. To do this, we simply note that, because of (14), $e_{jk}^{(\mu)} e_{kl}^{(\mu)} e_{lm}^{(\mu)} = e_{jm}^{(\mu)}$; on the other hand, if $e_{jk}^{(\mu)}$ and $e_{lm}^{(\nu)}$ belong to two different irreducible representations, then, because of (14) and the fact that every x in $\mathfrak{U}$ can be expanded in the e's, $e_{jk}^{(\mu)} x e_{lm}^{(\nu)} = 0$ for all $x \in \mathfrak{U}$. A special case of this, the only one we will need, is

Theorem 3: If e_1 and e_2 are two primitive idempotents, a necessary and sufficient condition for their belonging to the same IR is the existence of an $x \in \mathfrak{U}$ such that $e_1 x e_2 \neq 0$.

C. Induced and Subduced Representations

1. Subduced Representations

Let $\mathfrak{H}$ be a subgroup of $\mathfrak{G}$, and consider a representation Γ of $\mathfrak{G}$: $s \to D(s)$. It is evident that the $D(u)$ for $u \in \mathfrak{H}$ provide a representation γ of $\mathfrak{H}$, which is called the representation of $\mathfrak{H}$ subduced by Γ. We use the notation

$$\gamma(\mathfrak{H}) \equiv \{\Gamma(\mathfrak{G})\}_\mathfrak{H}$$

for the subduced representation. In general, if Γ is irreducible, γ need not be.

2. Induced Representations

In this subsection, the theory of the induced representation will be formulated in a manner similar to the treatment of Ruch and Schönhofer [13]. For a more standard mathematical treatment, see Boerner [9].

Again, suppose we have a group $\mathfrak{G}$ with a subgroup $\mathfrak{H}$. Let $\gamma(\mathfrak{H})$: $u \to d(u)$ be a representation of $\mathfrak{H}$, of dimension f, operating in a space r with basis vectors $|j>$, $j = 1, 2, \ldots, f$. Suppose also that r is a subspace of a larger space R, in which all the operations of $\mathfrak{G}$ are defined.

We form the left cosets of $\mathfrak{H}$:

$$t_1 \mathfrak{H} = \mathfrak{H} \; ; \; t_2 \mathfrak{H} \; ; \; \ldots \; ; \; t_q \mathfrak{H}, \text{ with } q = \frac{g}{h}.$$

Since all operations of $\mathfrak{G}$ are defined in R, we can generate the subspaces

$$r_\alpha = t_\alpha r,$$

each one of which is spanned by the basis vectors

$$|\alpha j> = t_\alpha |j>.$$

These q subspaces may or may not be independent. It is easily seen that all vectors $s|j>$, with $s \in \mathfrak{G}$, are linear combinations of the $|\alpha j>$; for, since

s must belong to one of the left cosets of $\mathfrak{H}$, we can write $s = t_\alpha u$, $u \in \mathfrak{H}$. It follows that

$$s|j\rangle = t_\alpha u|j\rangle = t_\alpha \sum_k |k\rangle \langle k|d(u)|j\rangle = \sum_k |\alpha k\rangle \langle k|d(u)|j\rangle.$$

It is also easy to see that each r_α forms a basis space for the subgroup $t_\alpha \mathfrak{H} t_\alpha^{-1}$. For, if $u \in \mathfrak{H}$

$$t_\alpha u t_\alpha^{-1}|\alpha j\rangle = t_\alpha u|j\rangle = t_\alpha \sum_k |k\rangle \langle k|d(u)|j\rangle = \sum_k |\alpha k\rangle \langle k|d(u)|j\rangle.$$

The space spanned by all the $|j\rangle$, $|\alpha j\rangle$ forms the basis for a representation of $\mathfrak{G}$ which we call the representation induced by $\gamma(\mathfrak{H})$. To see the form of the matrices of this representation, consider an element $s \in \mathfrak{G}$ applied to an arbitrary basis vector $|\alpha j\rangle$:

$$s|\alpha j\rangle = st_\alpha |j\rangle.$$

The element st_α, being in $\mathfrak{G}$, must be in one of the left cosets of $\mathfrak{H}$:

$$st_\alpha = t_\beta u, \quad u \in \mathfrak{H}.$$

We thus find

$$s|\alpha j\rangle = t_\beta u|j\rangle = t_\beta \sum_k |k\rangle \langle k|d(u)|j\rangle = \sum_k |\beta k\rangle \langle k|d(u)j\rangle. \quad (29)$$

If the subspaces r_α are all independent, the induced representation has dimension $F = fq$, and we speak of the induction as "regular" [13]. In the mathematical literature, non-regular induction is normally not discussed, and the word "induction" is used for what we call regular induction [9]. We will use the notation

$$\Gamma(\mathfrak{G}) = [\gamma(\mathfrak{H})]^{\mathfrak{G}}$$

for the regularly induced representation. According to (29), the matrix elements for the regularly induced representation are given by

$$\langle \beta k|D(s)|\alpha j\rangle = \langle k|d(t_\beta^{-1} st_\alpha)|j\rangle \quad \text{if } t_\beta^{-1} st_\alpha \in \mathfrak{H};$$
$$= 0 \text{ otherwise.}$$

In the case of non-regular induction, one would have to express each $|\alpha j\rangle$ in terms of a complete set for the induced representation space and then apply (29) to get the matrix elements. For general induction, not necessarily regular, it is evident that the dimension F of the induced representation satisfies

$$f \leqslant F \leqslant qf.$$

Before proving the key theorem on induced representations, it may help the reader to orient himself if we give a simple example of induction, both regular and non-regular. Consider a methane molecule, with symmetry T_d, and choose a coordinate system with z-axis in the direction of the bond from the central carbon atom (C) to hydrogen atom number 1 (H_1). The subgroup of T_d which leaves H_1 invariant is C_{3v}, of order 6, with $q=4$. Under this subgroup, the p_x and p_y orbitals on C transform between themselves according to the two-dimensional representation E of C_{3v}. When the other operations from T_d are applied to these two orbitals, one generates the space spanned by all three p-orbitals of C. This gives a representation of T_d, the three-dimensional one normally called F_2. In our language, we would say that the representation F_2 of T_d has been induced by E of C_{3v}. In this case, the induction is not regular.

Now consider the bending vibrations (in the xy plane) of H_1. They also transform among themselves according to E under C_{3v}. When the operations of T_d are applied to them, however, we generate the 8-dimensional space of all motions of the H-atoms perpendicular to the bonds (all bending vibrations, plus rotation of the molecule). The dimension of the representation of T_d so generated has dimension $8=qf$, so this time the induction is regular. When broken into its irreducible components, this regularly induced representation contains the two-dimensional representation E of T_d, as well as the two three-dimensional ones F_1 and F_2.

We now consider the representation Γ of $\mathfrak{G}$ induced by the irreducible representation γ of $\mathfrak{H}$. We pose the question: when Γ is broken up into its irreducible parts, how many times will each irreducible representation $\Gamma^{(\mu)}$ appear? To decide this, we first observe that the independent $|\alpha j\rangle$ which form a basis for Γ can all be generated from a single $|j\rangle$ of r by application of the elements of $\mathfrak{G}$, including those of $\mathfrak{H}$. For the vectors of r, this follows from the irreducibility of γ, for the others from the nature of the induction process. Since the e-operators form a complete set in $\mathfrak{U}$, one can equally well say that the basis vectors of Γ are generated by applying all the e-operators to a single $|j\rangle$. Now suppose that γ appears c times in the subduced representation $\{\Gamma^{(\mu)}(\mathfrak{G})\}_{\mathfrak{H}}$. Choose a basis for $\Gamma^{(\mu)}$ such that the first c basis vectors transform like $|j\rangle$ under $\mathfrak{H}$. In this basis we have

$$e^{(\mu)}_{kl} |j\rangle = 0, l > c.$$

For $l \leqslant c$, $e_{kl}^{(\mu)} |j\rangle$ may be zero, but need not be. It follows that we can generate from $|j\rangle$ at most c vectors transforming like the k'th basis vector of $\Gamma^{(\mu)}$. When the operations of $\mathfrak{G}$ are applied to these, each may separately generate the representation $\Gamma^{(\mu)}$, but it cannot be generated more times than this. Thus, the representation $\Gamma^{(\mu)}$ appears in Γ at most the same number of times that γ appears in the subduced representation of $\Gamma^{(\mu)}$ in $\mathfrak{H}$. We express this result more formally as follows. If $\Gamma(\mathfrak{G})$ is induced by the irreducible $\gamma^{(\sigma)}(\mathfrak{H})$, and its decomposition into irreducible components is

$$\Gamma = \sum a_\mu \, \Gamma^{(\mu)},$$

then we have

$$a_\mu \leqslant c_{\mu\sigma}, \tag{30}$$

where

$$\{\Gamma^{(\mu)}(\mathfrak{G})\}_{\mathfrak{H}} = \sum_\sigma c_{\mu\sigma} \, \gamma^{(\sigma)}. \tag{31}$$

We now show that the equality in (30) applies in the case of regular induction. Expressed in terms of characters (and using ξ to denote characters for $\mathfrak{H}$, (31) reads

$$\chi^{(\mu)}(u) = \sum_\sigma c_{\mu\sigma} \, \xi^{(\sigma)}(u).$$

We now multiply by $\xi^{(\tau)*}(u)$, sum over $u \in \mathfrak{H}$, use (3), and take the complex conjugate of the result to obtain

$$\sum_{u \in \mathfrak{H}} \xi^{(\tau)}(u) \, \chi^{(\mu)*}(u) = h \, c_{\mu\tau}. \tag{32}$$

Next multiply (32) by $n_\mu = \chi^{(\mu)}(1)$, sum over μ, and use (5), plus the fact that $\xi^{(\tau)}(1) = n_\tau$. This gives

$$g n_\tau = h \sum c_{\mu\tau} \, n_\mu,$$

or

$$\sum c_{\mu\tau} \, n_\mu = \frac{g}{h} \, n_\tau. \tag{33}$$

The right-hand side of (33) is just the dimension of the regularly induced representation $[\gamma^{(\tau)}(\mathfrak{H})]^{\mathfrak{G}}$, while the left-hand side is the dimension of the induced representation if the equality is satisfied in (29). This establishes our result, which may be written as

$$[\gamma^{(\tau)}(\mathfrak{H})]^{\mathfrak{G}} = \sum c_{\mu\tau} \, \Gamma^{(\mu)} \tag{34}$$

with the $c_{\mu\tau}$ defined by (31). The reader may wish to check this result for the case of the example given above.

D. Symmetric and Hyperoctahedral Groups

1. Definitions and General Remarks

In this article, we shall always be concerned with permutations of ligands among sites belonging to a molecular skeleton. No real generality will be lost if we think of permutations in this way from the outset; indeed, this differs only in the mental picture invoked from formulations commonly used by mathematicians, as when Weyl [14] speaks of permuting "men" among "fields" on a checkerboard.

We consider, then, a skeleton with n numbered sites to which ligands may be attached. If the ligands themselves are structureless, all isomers of a given molecule are generated by permuting the ligands among the sites. The group of all such permutations is the symmetric permutation group, $\mathfrak{S}_n$, of order $n!$. In our formulation, a permutation s belonging to $\mathfrak{S}_n$ will always be referred to *sites*: thus, s is defined by associating to each site j another one s_j (or $s(j)$ if the notation for a site is too complicated to be put in a subscript), in such a way that no two different j's have the same s_j. The effect of s, therefore, is to take the ligand initially found at site j and remove it to site s_j, regardless of which ligand that may be. It does *not* mean that one takes the j'th ligand and puts it where the s_j'th formerly was. In referring our permutations to sites rather than ligands, we are using the formulation of Weyl [14], which we feel is better suited for our purposes. The other formulation is also possible, of course, and is used by Boerner [9] and others.

If the ligands are allowed to be chiral, but are otherwise structureless, we can generate isomers not only by permutations but also by "site reflections" τ_j, which means: "replace the ligand on site j by its mirror image". The group of all combinations of permutations and site reflections is of order $2^n n!$ ($n!$ permutations, followed by either reflecting or not on each site), and is called the hyperoctahedral group [10,15] $\bar{\mathfrak{S}}_n$. A useful subgroup of $\bar{\mathfrak{S}}_n$, which itself contains $\mathfrak{S}_n$ as a subgroup is

$$\mathfrak{S}_n^! = \mathfrak{S}_n \times (1, \tau_0),$$

where

$$\tau_0 = \prod_{j=1}^{n} \tau_j$$

is the operation of simultaneous site reflection at all sites. $\mathfrak{S}_n^!$ is thus a direct product group of order $2n!$.

Returning to $\bar{\mathfrak{S}}_n$, an element σ of it can be expressed as a product τs, where s is a permutation and τ is some combination of reflections. Moreover, since τ operators do not permute the ligands, the permutation part of σ is independent of whether we write s to the left or the right:

$$\sigma = s\tau = \tau' s \tag{35}$$

with $\tau' = s\tau s^{-1}$. Now, if $\tau = \Pi \tau_j$, the product going over some set of sites, then

$$\tau' = s\Pi \tau_j s^{-1} .$$

Consider the fate of the ligand initially at the site s_j under the above operator. It is first carried by s^{-1} to the site j, where it is reflected by τ_j, then carried back to s_j. We conclude that

$$\tau' = s\Pi \tau_j s^{-1} = \Pi \tau(s_j) \equiv \tau^{(s)} . \tag{36}$$

The group $\bar{\mathfrak{S}}_n$ is a "semidirect product" group

$$\bar{\mathfrak{S}}_n = \mathfrak{S}_n \vee \mathfrak{T} ,$$

where $\mathfrak{T}$ is the group of the site reflections without permutation. $\mathfrak{T}$ is an invariant subgroup since, according to (36), $s\tau s^{-1}$ is in $\mathfrak{T}$. $\mathfrak{S}_n$, however, is not invariant, since

$$\tau^{-1} s \tau = \tau^{-1} \tau^{(s)} s$$

is not necessarily in $\mathfrak{S}_n$.

We shall be applying permutations and reflections to molecules to give other molecules (isomers), but also to functions of the properties of ligands on the sites. A simple example will suffice to show how permutations may be interpreted in this way.

Consider a skeleton with three sites. A molecule M is specified by specifying the ligand on each site. An example of a function for the class of molecules belonging to this skeleton is

$$F(M) = f(1)g(2)h(3) .$$

Thus, to calculate the property described by F, one takes the product of the function f for the ligand on site 1, g for the one on site 2, and h for the one on site 3. For example, if M_0 is a molecule with ligand a on site 1, b on 2, c on 3, the property F for M_0 has the value

$$F(M_0) = f(a)g(b)h(c) .$$

If we apply a permutation to M_0, giving a new isomer, the property F for the isomer will in general be different. Thus, if $s_1 = 2$, $s_2 = 3$, $s_3 = 1$, then $M_1 = sM_0$ will have ligand a on site 2, b on 3, and c on 1. For this isomer, the property F is

$$F(sM_0) = F(M_1) = f(c)g(a)h(b) \ .$$

Another viewpoint, however, is to consider a permutation as permuting, not the ligands, but the roles played by the sites in a function. From this point of view, the permutation s^{-1} (taking 1 to 3, 3 to 2, 2 to 1) defines a new function in which the role previously played by site 1 has been transferred to site 3, etc. In our example, this means

$$\overline{F}(M) = s^{-1}F(M) = f(3)g(1)h(2) \ .$$

For the molecule M_0, $\overline{F}$ has the value

$$\overline{F}(M_0) = s^{-1}F(M_0) = f(c)g(a)h(b) = F(sM_0) \ .$$

It is easy to convince oneself that this result is general; a permutation can be thought of as giving a new molecule sM by permuting the ligands, with property-describing functions naturally having altered values, or alternatively as permuting functional roles to give a new property (function) of the same molecule. The two are related by

$$s^{-1}F(M) = F(sM) \ .$$

2. Classes

We consider first the class structure of $\mathfrak{S}_n$. To do this, we note that every permutation may be written as the product of a number of independent cyclic permutations. Thus, the permutation s takes the ligand on site 1 to s_1, that on s_1 to $s(s_1)$, that on $s(s_1)$ to $s[s(s_1)]$, etc. Following this chain, since n is finite, we must eventually reach a site whose ligand is taken to site 1 by s. The closed chain evidently forms a cyclic permutation, which we will denote by writing the sites concerned in order enclosed in parentheses. Thus, $(123\ldots f)$ denotes a permutation s for which $s_1 = 2$, $s_2 = 3$, $\ldots s_{f-1} = f$, $s_f = 1$. If this first cycle does not include all the sites, we can do the same thing with the lowest-numbered site not appearing in the cycle, and continue until we have broken down the group element completely into a product of cycles.[b]

---

[b] In cases where this will cause no confusion, the "one-cycles" (cycles consisting of a single site, which is thus left undisturbed) will sometimes not be explicitly written down.

Now consider a group element r written in cyclic form:

$$r = (123\ldots f)(f+1, f+2, \ldots g)\ldots,$$

and conjugate it with another element s, forming srs^{-1}. Under srs^{-1}, the ligand on site s_1 is first taken to 1 by s^{-1}, then to 2 by r, then to s_2 by s; the ligand on s_2 is taken to 2 by s^{-1}, to 3 by r, to s_3 by s, etc. We see, then, that

$$srs^{-1} = (s_1 s_2 s_3 \ldots s_f)(s_{f+1}\, s_{f+2} \ldots s_g)\ldots$$

Since our r and s were essentially arbitrary, it follows that two elements belong to the same class if and only if they have the same "cycle structure", that is, if each consists of the same number of cycles of each length. We denote a class by listing the lengths of the cycles, with exponents to indicate when there are two or more cycles of the same length. For example, the classes of $\mathfrak{S}_5$ in this notation are:

$$(1^5);\ (1^3, 2);\ (1^2, 3);\ (1, 2^2);\ (1, 4);\ (2, 3);\ (5),$$

where, for example, the third class listed is that consisting of all elements made up of two 1-cycles and one 3-cycle; in other words, each permutation of this class leaves two sites invariant and effects a cyclic permutation of the other three. Evidently, there is a class for every way of writing n as a sum of integers (including $n = n$, $n = 1 + 1 + \ldots + 1$).

A *transposition* is the interchange of ligands on two sites, leaving all other sites and ligands undisturbed. It is well-known that any permutation can be expressed as a product of transpositions. A product of an odd (even) number of transpositions is called an odd (even) permutation. Although the expression of a given permutation as a product of transposition is not always unique, the odd or even character is the same for all such decompositions.

The classes of $\mathfrak{S}_n^!$ are now easily worked out. Since it is a direct product group, its classes are also direct products. Obviously, each class c of $\mathfrak{S}_n$ yields two classes of $\mathfrak{S}_n^!$, c and $\tau_0 c$.

We now take up the class structure of $\bar{\mathfrak{S}}_n$, first considering conjugation of an arbitrary element $\sigma = \tau s$ with a pure permutation, r. We find

$$r\sigma r^{-1} = r\tau s r^{-1} = \tau^{(r)} r s r^{-1}.$$

We know from the treatment of $\mathfrak{S}_n$ given above that, if $s = (12\ldots f)(f+1,\ldots g)\ldots$, then

$$rsr^{-1} = (r_1 r_2 \ldots r_f)(r_{f+1} \ldots r_g)\ldots$$

This, together with the definition of $\tau^{(r)}$, tells us that this conjugation just replaces every site y by r_y, as regards both permutation and reflection properties. Thus, cycle structure is preserved, as well as the location of reflections in cycles.

To consider conjugation of σ with a reflection operator $\bar{\tau}$, we first rearrange σ somewhat. We write s in cyclic form, and then make use of the fact that τ is a product of several single site reflections τ_j. Each τ_j, of course, commutes with the other reflections, and also with all permutations which do not affect site j. We can, therefore, move each τ_j to the right through the cyclic factors of s until it reaches the cycle containing site j. In this way, σ is written as a product of independent factors, each of which consists of a cyclic permutation followed by a product of site reflections referring only to sites in the cycle. When we conjugate with $\bar{\tau}$, we can do the same thing with the τ_k making up $\bar{\tau}$ and $\bar{\tau}^{-1} = \bar{\tau}$. Each τ_k on the left is moved to the right until it reaches the cycle containing k, and each one on the right is moved to the left until it reaches the same cycle. The conjugate element $\bar{\tau}\sigma\bar{\tau}^{-1} = \bar{\tau}\tau s \bar{\tau}^{-1}$, therefore, breaks up into a product of factors of the form

$$\prod_k \tau_k \, \tau c \, \prod_k \tau_k ,$$

where c is a cyclic permutation, τ contains reflections only on sites contained in c, and the k also refer exclusively to sites contained in c. Because $\prod_k \tau_k$ is a commutative product, we can consider the conjugation as conjugating consecutively in any order with the individual k. Also, since the numbering of the sites is arbitrary, we can assume $c = (12\ldots f)$. For a single k, we have

$$\tau_k \tau c \tau_k = \tau_k \tau (12\ldots f)\tau_k = \tau_k \tau \, \tau_{k+1}(12\ldots f) = \tau_k \tau_{k+1} \tau c,$$

where the addition of 1 is modulo f, i.e., by definition $f + 1 = 1$. Conjugation with a single τ_k, therefore, has the effect of reflecting on two sites which are nearest neighbors in the cycle. Subsequent conjugation with τ_{k+1}, τ_{k+2}, etc., enables us to reflect on any two sites in the cycle by conjugation, and by further conjugation we can reflect on two other sites, then on two others, etc. Thus, by conjugating with reflection operators, we can bring about reflections on any *even* number of sites in the cycle. The conjugated element can differ from the original one in the distribution of reflections in any way so long as the difference in the total number of reflections is an even number. Thus, elements with an odd number of reflections in the cycle are all in the same class, those with an even number all in a separate class. Since any conjugation can be thought of as conjugating first with a permutation and then with a reflection operator, we conclude that a class of $\bar{\mathfrak{S}}_n$ is specified by giving the lengths of all cycles containing an even number of reflected sites (denoted as in $\mathfrak{S}_n$ by a number q giving the length of the cycle), as well as the lengths of all cycles containing an odd number of reflections (denoted by $\bar{q}$). For example, the classes of $\bar{\mathfrak{S}}_3$ are:

$$(1^3);\ (1,2);\ (3);\ (1^2,\bar{1});\ (1,\bar{2});\ (2,\bar{1});\ (\bar{3});\ (1,\bar{1}^2);\ (\bar{1},\bar{2});\ (\bar{1}^3).$$

The fifth class listed above contains a 1-cycle and a 2-cycle in its permutation part, with an odd number of reflections (in this case, one reflection) on sites involved in the two-cycle. The seventh class has a permutation part consisting of a cyclic permutation of the three sites, with an odd number of reflections (either one or three).

3. Representation Theory

Our treatment of the representation theory of $\mathfrak{S}_n$ in this subsection will be taken mostly from the books by Boerner [9] and Weyl [14]. The treatment of $\bar{\mathfrak{S}}_n$ will closely parallel that of Ref [7]. We first consider the case of $\mathfrak{S}_n$.

The representation theory of $\mathfrak{S}_2$ is, of course, trivial: There are two irreducible representations, the symmetric and the antisymmetric. For $\mathfrak{S}_n$, we also know that, among the irreducible representations, there is a totally symmetric one and a totally antisymmetric one. It is thus reasonable to assume, or at least to hope, that the other irreducible representations of $\mathfrak{S}_n$ can be obtained by some combination of symmetrizing and antisymmetrizing. We attempt to construct a combined symmetrizing-antisymmetrizing operator as follows:

We express n as a sum of integers $n = \mu_1 + \mu_2 + \ldots$, with $\mu_1 \geq \mu_2 \geq \mu_3 \geq \ldots$. Such a "partition" of n can be expressed, as illustrated in Fig. 1, by a "Young diagram of order n", in which n boxes are arranged in columns of length (from left to right) μ_1, μ_2, etc. Fig. 1(a) shows a Young diagram for $n=6$, with $\mu_1=3$, $\mu_2=2$, $\mu_3=1$. Fig. 1(b) is a more complicated diagram, for $\mathfrak{S}_{20}$, with $\mu_1=5$, $\mu_2=\mu_3=4$, $\mu_4=\mu_5=3$,

$\mu_6 = 1$. It is evident that a diagram may equally well be specified by the row lengths ν_1, ν_2, etc. (For Fig. 1(a), these are 3,2,1; for Fig. 1(b), they are 6,5,5,3,1). Also, since a class of $\mathfrak{S}_n$ is specified by a partition of n, it is clear that the number of different Young diagrams of order n is equal to the number of classes, hence to the number of irreducible representations of $\mathfrak{S}_n$. A Young diagram will be denoted by γ, sometimes with a superscript. Sometimes, instead of drawing a diagram, we will denote it by listing the lengths of its rows, with row lengths appearing more than once denoted by exponents. In this notation, the diagram of Fig. 1(a) is (3,2,1); that of Fig. 1(b) is $(6,5^2,3,1)$. An array of boxes in which any row is longer than the one above it is not a Young diagram.

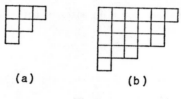

Fig. 1

From a given Young diagram we can, in $n!$ different ways, form a "Young tableau" t by filling the boxes, in any order, with the site numbers $1, 2, \ldots, n$. Fig. 2 shows examples of tableaux formed from the diagrams of Fig. 1.

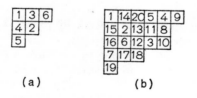

Fig. 2

A tableau may be used to define certain subgroups of $\mathfrak{S}_n$ which are themselves direct products of smaller permutation groups; the symmetrizing and antisymmetrizing operators for these subgroups lead, as we shall see, to projection operators on irreducible representations of $\mathfrak{S}_n$.

Given a tableau t, we first define the subgroup $\mathfrak{Q}$, consisting of all permutations q among sites in the same column of t. Evidently, $\mathfrak{Q}$ is just the direct product of the permutation groups for the sites in the various columns: $\mathfrak{Q} = \mathfrak{S}(\mu_1) \times \mathfrak{S}(\mu_2) \times \ldots$. The antisymmetrizer for the subgroup is

$$Q = \sum_q \varepsilon_q q ,\qquad(37)$$

where $\varepsilon_q = \pm 1$, according as q is an even or an odd permutation. Having antisymmetrized with respect to certain groups of sites, we are still free to symmetrize with respect to certain others. We cannot, however, symmetrize with respect to sites in the same column; this, together with the antisymmetrization already carried out, would simply give zero. We must, therefore, choose sites from different columns, and without loss of generality we can consider these to be in the same row. Accordingly, we define, analogously to $\mathfrak{Q}$, another subgroup $\mathfrak{P}$, consisting of all permutations p among sites in the same row. The symmetrizer for this group is

$$P = \sum_p p .\qquad(38)$$

Combining the operators P, Q, we obtain the „Young operator" associated with the tableau t:

$$Y = PQ = \sum \varepsilon_q pq .\qquad(39)$$

(It should be noted that, since P does not in general commute with Q, a function $Yf = PQf$ is not necessarily still totally antisymmetric under $\mathfrak{Q}$.)

We will now proceed to prove

Theorem 1: The operator Y defined by (39) is, apart from a multiplicative constant, a primitive idempotent of $\mathfrak{S}_n$. Y operators belonging to the same Young diagram belong to the same irreducible representation, while those belonging to different diagrams belong to different representations.

The first part of the theorem will be proved by showing that the Y operators satisfy the conditions of Theorem 2, Subsection B. The second part will be obtained by applying the criterion of Theorem 3 of that subsection. The proof proceeds by way of several intermediate theorems, as follows:

i) For arbitrary p, q, we have

$$pP = P, \quad Qq = \varepsilon_q Q ;\qquad(40)$$

$$pYq = \varepsilon_q Y .\qquad(41)$$

Also, for arbitrary $x \in \mathfrak{U}$, p, q,

$$p(YxY)q = \varepsilon_q(YxY) \,. \tag{42}$$

The results (40) follow immediately from the definitions (37), (38); (41) follows from the definition (39) and application of (40); (42) is obtained by twice applying (41).

ii) Two ligands initially on sites in the same column of t are never moved into sites in the same row of t by a permutation pq; and all permutations having this property can be written in the form pq.

This can be seen essentially by inspection: After a permutation q, ligands initially in the same column are still in the same column, hence in different rows; and the permutation p does not move a ligand from one row to another. On the other hand, to bring about a permutation in which ligands from the same column remain in different rows, we can first put each ligand into the row of its destination by a q permutation, then reach the final arrangement by means of a p.

iii) Let t, t' be two tableaux belonging to the same diagram, and let s be the permutation with the property that s_j has the same location in t' as j has in t. Let $u(t)$ be an arbitrary permutation, and $u(t')$ be the permutation obtained from u by replacing each site j by s_j, the site playing the same role in t' as j does in t. Then

$$su(t)s^{-1} = u(t'),$$

and in particular

$$sY(t)s^{-1} = Y(t') \,. \tag{43}$$

Proof: Under $su(t)s^{-1}$, the ligand initially on the site s_j is taken by s^{-1} to j, then by $u(t)$ to u_j, then by s to $s(u_j)$, which was the result to be proved.

iv) If a permutation $r \neq pq$, then there is a transposition p and a transposition q such that $prq = r$.

Proof: Because of (ii) above, there must be two sites j,k in the same column of t, such that r_j, r_k are in the same row of t. Let $q = (jk)$, $p = (r_j r_k)$. Then

$$prq = (r_j r_k)r(jk) = r(jk)r^{-1}r(jk) = r(jk)(jk) = r.$$

v) If $a \in \mathfrak{U}$ has the property that

$$paq = \varepsilon_q a \quad \text{for arbitrary } p, q, \text{ then}$$
$$a = fY, \quad \text{where } f \text{ is a constant.}$$

Proof: Writing a in the form

$$a = \sum_s a(s)s ,$$

we find that the assumed property reads

$$\sum_s a(s) psq = \varepsilon_q \sum_s a(s)s. \tag{44}$$

The permutation pq itself appears on the left side of (44) for $s=1$, on the right side for $s=pq$. This gives

$$\varepsilon_q a(pq) = a(1) ,$$
$$a(pq) = \varepsilon_q q(1) . \tag{45}$$

Now, for $r \neq pq$, we choose the transpositions p and q such that $prq = r$, whose existence was established in (iv) above. Since this q is a transposition, we have $\varepsilon_q = -1$. Since $prq = r$, r appears on both the left and right sides of (44) for $s = r$, and we have

$$a(r) = \varepsilon_q a(r) = -a(r) ,$$
$$a(r) = 0. \tag{46}$$

Because of (46), if a is to have the assumed property it must consist only of permutations pq. Because of (45), the coefficients of these must be, apart from an overall multiplicative constant, just ε_q. This completes the proof.

vi) Y is a primitive idempotent, up to a multiplicative constant.

Proof: We obviously have $Y(1) = 1 \neq 0$. Because of (42), we see that (YxY) for arbitrary $x \in \mathfrak{U}$ satisfies the condition of (v) above, so it follows that $YxY = fY$, for arbitrary $x \in \mathfrak{U}$. Thus, the requirements of Theorem 2, Subsection B, are fully satisfied.

vii) Y operators belonging to the same diagram γ belong to the same IR; those belonging to different diagrams belong to different IR's.

Proof: If $Y = Y(t)$, $Y' = Y(t')$ arise from tableaux belonging to the same diagram, then, according to (iii), we have

$$Y' = sYs^{-1} , \text{ and}$$

$$Y'sY = sYs^{-1}sY = sY^2 = sfY \neq 0 .$$

The last inequality follows from the fact that, if $sY=0$, we would have $s^{-1}sY=Y=0$. Thus, according to the criterion of Theorem 3, Subsection B, Y and Y' belong to the same IR.

Now suppose that Y, Y' belong to different diagrams, γ and γ' with the row lengths $v_j = v'_j$ for the first m rows, $j \leqslant m$, but with the $(m+1)$'st row being longer in γ' than in γ. It follows that there must be at least two sites in the same row in t' which are in the same column in t. For, if we try to place the numbers from the first row of t' in the boxes of γ with no two in the same column, we will have to put one in each column of γ. Doing the same with those from the second row, we will have to put one in each column which still has vacancies, etc. After transferring the first m rows in this way, by hypothesis the number of columns still having vacancies will be less than the number of sites in the next row to be transferred, so at least two sites from this row must go into the same column of γ. If j,k are two sites in the same row of t' and the same column of t, then, since P' symmetrizes, and Q antisymmetrizes, with respect to the transposition (jk), we have

$$QP' = 0, \quad YY' = PQP'Q' = 0$$

Also, from (43), $sY's^{-1} = Y''$ for any permutation s, where Y'' is a Y operator belonging to γ'. We thus have, for any s,

$$YsY' = YY''s = 0,$$

since $YY' = 0$ by the same reasoning as above. For arbitrary $y \in \mathfrak{U}$, we have

$$YyY' = \sum_s y(s) YsY' = 0.$$

Thus, according to the criterion of Theorem 2, Subsection B, Y, Y' belong to different IR's. This completes the proof of Theorem 1, stated above. By means of Theorem 1, we obtain a one-one correspondence (via the Young operators) between diagrams γ and IR's Γ: $\gamma \rightleftarrows \Gamma$. We will sometimes refer to a representation and its diagram interchangeably.

We will also make some use of the fact that $Y^* = QP$ is also a constant times a primitive idempotent on the same IR as Y. That Y^* is a constant times a primitive idempotent can be proved analogously to Theorem 1. That it belongs to the same IR as Y follows from

$$Y^*Y = QPPQ = fQPQ \neq 0.$$

(If $QPQ = 0$, it would follow that $PQPQ = Y^2 = 0$.)

A „standard tableau" is defined as one in which the numbers increase when one reads from left to right in each row and from top to bottom in each column. It can be shown [9,11] that the dimension of a representation is equal to the number of standard tableaux associated with the corresponding diagram. The reader is also referred to the literature [9,11] for methods of calculating characters of the representations from the diagrams.

Since $\mathfrak{S}_n^!$ is a direct product group, its irreducible representations are also direct products. We denote a representation of $\mathfrak{S}_n^!$ by a Young diagram giving the representation of $\mathfrak{S}_n$, together with a letter g or u according as the representation is even or odd with respect to τ_0.

The representation theory of the groups $\bar{\mathfrak{S}}_n$ has been considered by Young [10] and Frame [15]. It can be developed in a diagrammatic manner quite analogous to that of $\mathfrak{S}_n$, as follows:

We define a "two-part diagram" $\bar{\gamma}$ of order n as a pair of ordinary Young diagrams $\gamma^{(+)}$ (order n_+, labelled with a $(+)$ sign), and $\gamma^{(-)}$ (order n_-, labelled with a $(-)$ sign), such that $n_+ + n_- = n$. An example with $n = 8$ is shown in Fig. 3. It is easy to see that the number of different two-part diagrams is equal to the number of classes, hence to the number of IR's of $\bar{\mathfrak{S}}_n$: For each class, one can construct a two-part diagram whose row lengths are those of the even cycles in the $(+)$ part, and of the odd cycles in the $(-)$ part. Clearly, this correspondence is one-to-one. A two-part diagram may be denoted by listing the row lengths of its parts, e.g., $(3,2+; 2,1-)$ for Fig. 3.

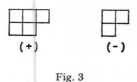

Fig. 3

By filling the boxes of a two-part diagram with the site numbers $1, 2, \ldots, n$ we arrive at a two-part tableau $\bar{t}$ of order n with parts $t^{(+)}$, $t^{(-)}$. Analogously to the case of $\mathfrak{S}_n$, we can use a tableau $\bar{t}$ to define a set of subgroups of $\bar{\mathfrak{S}}_n$ and symmetrizing and antisymmetrizing operators associated with them. Thus, for the $(+)$ part of the tableau we define the group of row permutations $\mathfrak{P}_+$ with members p_+, and the group of column permutations $\mathfrak{Q}_+$, with members q_+, exactly as for an ordinary diagram, and in the same way we define $\mathfrak{P}_-$, p_-, $\mathfrak{Q}_-$, q_- associated with the minus part. We will also need the direct product groups $\bar{\mathfrak{P}} = \mathfrak{P}_+ \times \mathfrak{P}_-$ (members $\bar{p}$); $\bar{\mathfrak{Q}} = \mathfrak{Q}_+ \times \mathfrak{Q}_-$ (members $\bar{q}$). We further define two new subgroups: $\mathfrak{T}_+$ (members τ_+) consisting of all reflection operators on sites in $t^{(+)}$, and $\mathfrak{T}_-$ (members τ_-) defined analogously for $t^{(-)}$. Analogously to (37), (38), we define a set of symmetrizing and antisymmetrizing operators as follows:

$$Q_+ = \sum \varepsilon(q_+)q_+ \; ; \; Q_- = \sum \varepsilon(q_-)q_- \; ; \; \overline{Q} = Q_+Q_- = \sum \varepsilon(\bar{q})\bar{q} \,. \quad (47)$$

$$P_+ = \sum p_+ \; ; \; P_- = \sum p_- \; ; \; \overline{P} = P_+P_- = \sum \bar{p} \,. \quad (48)$$

$$T_+ = \sum \tau_+ = \prod_{j \in t^{(+)}} (1 + \tau_j) \; ; \; T_- = \sum \varepsilon(\tau_-)\tau_- = \prod_{k \in t^{(-)}} (1 - \tau_k) \,, \quad (49)$$

where $\varepsilon(\tau_-) = +1$ if the number of reflections in τ_- is even, -1 if it is odd. Of the operators defined in Eqs. (47—49), the only noncommuting pairs are (P_+, Q_+), (P_-, Q_-) and the others obviously derivable from these such as $(\overline{P}, \overline{Q})$.

The Young operator for the tableau is defined as

$$\overline{Y} = T_+\overline{PQ}T_- = T_+Y_+Y_-T_- \,. \quad (50)$$

Because of the commutativity, $\overline{Y}$ may be written in a number of different ways, such as $T_+T_-\overline{PQ}$, etc. Analogously to Theorem 1 for ordinary Young operators, we have

Theorem 2: The operator $\overline{Y}$ defined by (50) is, apart from a multiplicative constant, a primitive idempotent for $\overline{\mathfrak{S}}_n$. $\overline{Y}$ operators belonging to the same two-part diagram belong to the same irreducible representation; those belonging to different diagrams belong to different representations.

The proof again proceeds in steps, which are quite analogous to those for Theorem 1. These are as follows:

i') Eq. (40) holds for P_+, P_-, $\overline{P}$; Q_+, Q_-, $\overline{Q}$. We also have

$$\tau_+T_+ = T_+ \; ; \quad (51)$$

$$\tau_-T_- = \varepsilon(\tau_-)T_- \,. \quad (52)$$

Eqs. (41) and (42) generalize to

$$\tau_+\bar{p}\overline{Y}\bar{q}\,\tau_- = \varepsilon(\bar{q})\varepsilon(\tau_-)\overline{Y} \,, \quad (53)$$

and

$$\tau_+\bar{p}(\overline{Y}x\overline{Y})\bar{q}\tau_- = \varepsilon(\bar{q})\varepsilon(\tau_-)\,(\overline{Y}x\overline{Y}) \,. \quad (54)$$

All of these assertions follow immediately from the definitions.

ii') A permutation $\bar{p}q$ does not move any ligand from a site in one part of $\tilde{t}$ to a site in the other; nor does it move two ligands initially in the same column of $t^{(+)}$ or $t^{(-)}$ into the same row. All permutations having this property are expressible in the form $\bar{p}q$.

The proof of this is exactly analogous to that for (ii) above.

iii') The analogue of (43) is

$$\sigma \bar{Y}(\tilde{t})\sigma^{-1} = \bar{Y}(\tilde{t}') \,, \qquad (55)$$

where $\sigma = \tau s$, and $\tilde{t}'$ is the tableau obtained from $\tilde{t}$ by replacing each site j by s_j. This follows from the reasoning used under (iii) above, plus the fact, easily provable from (52), (53), and the commutativity properties of the operators, that τ commutes with $\bar{Y}$.

iv') There are now two parts to this intermediate theorem, viz:

(a) If $\varrho = \tau r$, where r is a permutation which does not mix the $(+)$ and $(-)$ parts of $\tilde{t}$ and is not of the form $\bar{p}q$, then there exist $\tau_+ \bar{p}$, $\tau_- \bar{q}$, where $\varepsilon(\tau_-) = +1$, and $\bar{p}$, $\bar{q}$ are both transpositions, such that

$$\tau_+ \bar{p} \varrho \bar{q} \tau_- = \varrho \,. \qquad (56)$$

Proof: Since r does not mix the parts of $\tilde{t}$ we can write ϱ as $\varrho = v_+ r v_-$, where v_+, v_- are reflection operations confined respectively to the $(+)$ and $(-)$ parts of $\tilde{t}$. We now choose $\bar{p}$ and $\bar{q}$ to be the transpositions (whose existence has been proved in (iv) above) with the property $\bar{p} r \bar{q} = r$. Let $\tau_+ = v_+ v_+(\bar{p})$, $\tau_- = v_-(\bar{q}) v_-$. By substitution into Eq. (56), one verifies immediately that these operators have the asserted properties.

(b) If $\varrho = \tau r$, where r is a permutation which mixes the $(+)$ and $(-)$ parts of $\tilde{t}$, then there exist a τ_+ and a τ_-, with $\varepsilon(\tau_-) = -1$, with the property that $\tau_+ \varrho \tau_- = \varrho$.

Proof: For an r which mixes the two parts, there is at least one site j in $t^{(-)}$ with r_j in $t^{(+)}$. Let $\tau_+ = \tau(r_j)$, $\tau_- = \tau_j$. One verifies immediately by substitution that these operators have the stated property.

v') If an operator $a \in \mathfrak{U}$ has the property that

$$\tau_+ \bar{p} a \bar{q} \tau_- = \varepsilon(\bar{q}) \varepsilon(\tau_-) a$$

for all $\bar{p}$, $\bar{q}$, τ_+, τ_-, then $a = f\bar{Y}$, where f is a number.

Proof: The proof is so completely analogous to that of (v) above that there is no need to give it in detail. One writes $a = \sum a(\sigma)\sigma$. For σ

whose permutation part is $\overline{pq}$, one shows, exactly as in (v), that $a(\sigma) = \varepsilon(\bar{q})\varepsilon(\tau_-)a(1)$. For other σ, one uses the results of (iv') exactly as we previously used (iv) to show that $a(\sigma) = 0$.

vi') $\bar{Y}$ is a number times a primitive idempotent.

Proof: Exactly as in (vi), we simply note that the conditions of Theorem 2, Subsection B, are satisfied.

vii') $\bar{Y}$ operators belonging to the same diagram belong to the same IR. Those belonging to different diagrams belong to different IR's.

Proof: If two $\bar{Y}$ operators belong to the same diagram, we use (55) to show that $\bar{Y}'\sigma Y \neq 0$. If $\bar{Y}$, $\bar{Y}'$ belong to different diagrams with, say, $n_+ > n'_+$, then there is at least one site j which is in $t^{(+)}$ and in $t'^{(-)}$. The product $\bar{Y}'\bar{Y}$ then contains a factor $(1-\tau_j)(1+\tau_j) = 0$. If $n_+ = n'_+$, but the diagrams are nevertheless different, one shows as in the proof of (vii) that $\overline{YY'} = 0$. The rest of the proof is exactly the same as that of (vii).

E. Regular Induction from $\mathfrak{S}_n$ and $\mathfrak{S}'_n$ to $\tilde{\mathfrak{S}}$

In later sections of this paper, it will be necessary to carry out regular induction from a subgroup $\mathfrak{G}$ of $\mathfrak{S}'_n$ to $\tilde{\mathfrak{S}}_n$. In principle, the induction can be carried out in a straightforward way using the results of Section II-C, if character tables for $\tilde{\mathfrak{S}}'_n$ are available; moreover, such character tables are available at least in principle, as a formula exists for calculating them from the readily available ones of $\mathfrak{S}_n$ [7]. Even for relatively small n, however, this procedure is extremely cumbersome. The induction is much more conveniently and elegantly carried out in two steps, inducing first from $\mathfrak{G}$ to $\mathfrak{S}'_n$ and then from $\mathfrak{S}'_n$ to $\tilde{\mathfrak{S}}_n$. If the first step yields

$$[\Gamma(\mathfrak{G})]^{\mathfrak{S}'_n} = \sum_r c_r \Gamma^{(r)}(\mathfrak{S}'_n), \tag{57}$$

and the second step gives

$$[\Gamma^{(r)}(\mathfrak{S}'_n)]^{\tilde{\mathfrak{S}}_n} = \sum_s c_{rs} \Gamma^{(s)}(\tilde{\mathfrak{S}}_n), \tag{58}$$

then we evidently have for the final result

$$[\Gamma(\mathfrak{G})]^{\tilde{\mathfrak{S}}_n} = \sum_{rs} c_r c_{rs} \Gamma^{(s)}(\tilde{\mathfrak{S}}_n). \tag{59}$$

The first step, induction from $\mathfrak{G}$ to $\mathfrak{S}'_n$, is easily carried out with the aid of the character tables. The second step can be effected without the use of character tables, by means of a simple procedure which we now describe. We omit the rather lengthy proof, which is found in Ref[7]. We first describe the procedure for regular induction from $\mathfrak{S}_n$ to $\bar{\mathfrak{S}}_n$, then indicate the simple modification needed for $\mathfrak{S}'_u$. The representation $[\Gamma^{(r)}]^{\bar{\mathfrak{S}}_n}$ of $\bar{\mathfrak{S}}_n$ regularly induced by $\Gamma^{(r)}$ of $\mathfrak{S}_n$ is constructed as follows:

We fill some of the n boxes of the diagram $\gamma^{(r)}$ with $\varkappa_1$ symbols 1, $\varkappa_2$ symbols 2, ... $\varkappa_t$ symbols t subject to the conditions $\varkappa_1 \geq \varkappa_2 \geq \ldots \varkappa_t$ and, $\varkappa_1 + \varkappa_2 + \ldots + \varkappa_t \leq \dfrac{u}{2}$ in such a way that

1) No two symbols of the same kind are in the same column.

2) When all the boxes containing the symbol t are removed the remaining boxes constitute a diagram; this must remain so when all boxes with t and $t-1$ are removed etc.

3) Reading from right to left along the rows, beginning with the top row and working down, one must get a "lattice permutation" of 111... 22... That is, at each point one must have read at least as many 1's as 2's, as many 2's as 3's, etc.

For each way of doing this, the induced representation contains an irreducible representation whose (+) diagram is the diagram of empty boxes and whose (−) diagram has rows of lengths $\varkappa_1, \varkappa_2, \ldots, \varkappa_t$. If the plus and minus diagram are of different order, the induced representation also contains a second irreducible representation corresponding to a two-part diagram in which the plus and minus diagrams are interchanged, otherwise not[c].

The induction from $\mathfrak{S}'_n$ to $\bar{\mathfrak{S}}_n$ is the same, except that the only induced diagrams allowed are those with an even or odd number of boxes in the negative part according as the inducing representation is g or u.

In Fig. 4, we give an example of the induction from $\mathfrak{S}_7$ to $\bar{\mathfrak{S}}_7$.

For induction from $\mathfrak{S}'_7$ to $\bar{\mathfrak{S}}_7$, we must distinguish between the indices g or u of the inducing representation and consequently ignore induced diagram pairs with an odd (for g) or even (for u) number of boxes in the minus component. Thus, referring to Fig. 4, if the inducing representation of $\mathfrak{S}'_7$ is $(3,2^2)_u$, we obtain only the representations under column (u) in the figure; if it is $(3,2^2)_g$ we get only those in column (g).

---

[c] If the two parts are of the same order, but are not identical, there is always another allowed way of filling in the symbols which leads to the diagram pair with interchanged signs. The prescription as stated prevents this from being counted twice.

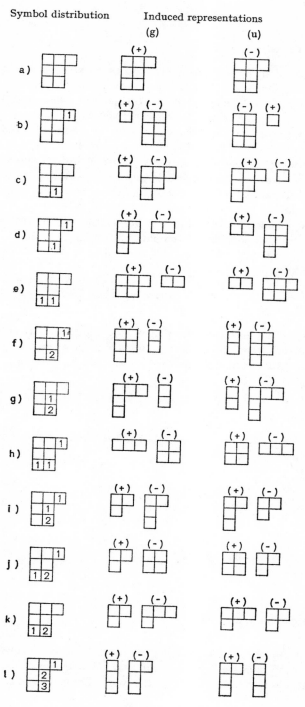

Fig. 4

F. The Transfer Condition

It will prove to be important later on to be able to decide whether a certain relation is satisfied between two Young diagrams of the same order. We call this the transfer condition or "T-condition". For one-part diagrams ($\mathfrak{S}_n$), this relation is defined as follows:

Given two diagrams of the same order, $\gamma^{(r)}$ and $\gamma^{(p)}$, we imagine the boxes of $\gamma^{(p)}$ to be filled with symbols 1 in the first row, 2 in the second, etc. We say that the T-condition is satisfied for $\gamma^{(p)}$ into $\gamma^{(r)}$ if it is possible to transfer these symbols into the boxes of $\gamma^{(r)}$ in such a way that no two like symbols are in the same column of $\gamma^{(r)}$ (Column condition of C-condition).

To determine under what conditions the T-condition is satisfied, we note that, in order to accomodate all the 1's, the number of columns in $\gamma^{(r)}$ must be at least as large as the length of the first row of $\gamma^{(p)}$, in other words we must have

$$\nu_1^{(r)} \geqslant \nu_1^{(p)},$$

where ν_j is the length of the j'th row of a diagram. If the 1's can be accomodated, then the unused columns plus the ones of length $\geqslant 2$ must be enough to accomodate all the 2's, *i.e.*, we must have

$$\nu_1^{(r)} + \nu_2^{(r)} \geq \nu_1^{(p)} + \nu_2^{(p)}.$$

Continuing in this way, we see that the transfer condition is satisfied if and only if

$$o_j^{(r)} \geqslant o_j^{(p)}, \text{ all } j, \tag{60}$$

where

$$o_j = \sum_{k=1}^{j} \nu_k.$$

If (60) is satisfied, we say that

$$\gamma^{(r)} \supset \gamma^{(p)} \tag{61}$$

($\gamma^{(r)}$ is greater than $\gamma^{(p)}$).

It is also quite easy to see that moving boxes downward in a diagram never increases any o_j, and that all diagrams smaller than the original one can be obtained in this way. Thus, we can also formulate the "⊃" relation thus: $\gamma^{(r)} \supset \gamma^{(p)}$ if and only if $\gamma^{(p)}$ is obtainable from $\gamma^{(r)}$ by moving boxes downward (without at any point producing an array of boxes which is not a diagram).

We note further that moving boxes downward in this way also entails moving them from right to left. This leads to the analog of (60) in terms of the column lengths; $\gamma^{(r)} \supset \gamma^{(p)}$ if and only if

$$u_j{}^{(r)} \leqslant u_j{}^{(p)}, \text{ all } j, \qquad (62)$$

where

$$u_j = \sum_{k=1}^{j} \mu_k,$$

and μ_j is the length of the j'th column.

We thus conclude that the T-condition for $\gamma^{(p)}$ into $\gamma^{(r)}$ is satisfied if and only if $\gamma^{(r)} \supset \gamma^{(p)}$. The "⊃" relation is determined equivalently by (60), (62), or the condition of downward movement of boxes.

One should note that the "⊃" relation gives only a partial ordering of the Young diagrams. There can exist pairs of diagrams γ, γ', such that $\gamma \not\supset \gamma'$ and also $\gamma' \not\supset \gamma$. Thus, for the diagram of Fig. 5 we have $\gamma_3 \supset \gamma_2$, $\gamma_3 \supset \gamma_1$, but neither $\gamma_1 \supset \gamma_2$ nor $\gamma_2 \supset \gamma_1$. The "⊃" relation is transitive: If $\gamma_a \supset \gamma_b$ and $\gamma_b \supset \gamma_c$, it follows that $\gamma_a \supset \gamma_c$. It can be shown[6] that the "⊃" relation defines a lattice.

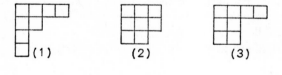

Fig. 5

The appropriate generalization of the T-condition for two-part diagrams is the following:

Consider two two-part diagrams of order n, $\bar{\gamma}^{(r)}$ and $\bar{\gamma}^{(p)}$, the boxes of the latter being filled with symbols 1 in the first row, 2 in the second, etc., with the symbols in the minus part distinguished by primes. The "generalized T-conditions" is said to be satisfied if all the symbols can be transferred into the boxes of $\bar{\gamma}^{(r)}$ in such a way that

1) No two like symbols appear in the same column of the same part of $\bar{\gamma}^{(r)}$ (primed and unprimed symbols being counted as different for this purpose); and

2) All the unprimed symbols are in the plus part of $\bar{\gamma}^{(r)}$ (though primed symbols may be in either part).

We now consider what properties the diagrams must have in order to satisfy this condition.

First, the plus part of $\bar{\gamma}^{(r)}$ must be able to accomodate all the unprimed symbols with no two like ones in the same column, though some boxes may be left unfilled. This means that the first row of $\bar{\gamma}^{(r+)}$ must be long enough to accomodate all the 1's, the leftover boxes in the first row plus those in the second row enough for the 2's, etc. Continuing in this way, by reasoning exactly analogous to that used above for single diagrams, one conludes that, to fulfill this part of the T-condition, it is necessary and sufficient that

$$o_i{}^{(r+)} \geqq o_i{}^{(p+)}, \text{ all } i.$$

Second, there must be enough different kinds of primed symbols to fill the first column of $\gamma^{(r-)}$, the unused ones plus those of which there are at least two must be enough to fill the second column, etc. Again, by reasoning analogous to the above, one concludes that this part of the T-condition is satisfied if and only if

$$u_i{}^{(r-)} \leqq u_i{}^{(p-)}, \text{ all } i.$$

The above two conditions are necessary, though not sufficient, for the T-condition. By means of them, we can define a "greater" relation for two-part diagrams which is analogous to the known one for ordinary diagrams, as follows:

We say that $\bar{\gamma}^{(r)} \supset \bar{\gamma}^{(p)}$ ($\bar{\gamma}^{(r)}$ greater than $\bar{\gamma}^{(p)}$) if

$$o_i{}^{(r+)} \geqq o_i{}^{(p+)}, \text{ all } i,$$

and

$$u_i{}^{(r-)} \leqq u_i{}^{(p-)}, \text{ all } i.$$

A second, equivalent, definition is: $\bar{\gamma}^{(r)}$ is greater than $\bar{\gamma}^{(p)}$ if and only if $\bar{\gamma}^{(p)}$ can be obtained from $\bar{\gamma}^{(r)}$ by moving boxes from the plus to the minus part, and/or downward within each part.

It is shown in Ref. [7] that this partial ordering defines a lattice.

If $\bar{\gamma}^{(r)} \supset \bar{\gamma}^{(p)}$, then the above considerations show that we can transfer all the unprimed symbols into $\gamma^{(r+)}$, with no two like ones in the same column and $n_{rp} = n^{(r+)} - n^{(p+)} = n^{(p-)} - n^{(r-)}$ boxes remaining empty. Also, we can fill $\gamma^{(r-)}$ with primed symbols, no two like ones in the same column, with n_{rp} symbols not being used. In general, this partial transfer can be done in several different ways. The T-condition now will be satisfied if we can transfer the n_{rp} remaining primed symbols into the n_{rp} empty boxes with no two like symbols going into the same column. We can express this condition as a T-condition for ordinary diagrams as follows: First, form a diagram $\gamma^{(rp+)}$ of order n_{rp} whose column lengths are the numbers of empty boxes in the columns of $\gamma^{(r+)}$. Form another diagram $\gamma^{(rp-)}$ of order n_{rp} whose row lengths are equal to the numbers of unused primed symbols of each kind. These diagrams will be called remainder diagrams. In terms of them, the question of transferability of the unused symbols becomes: "Can all the symbols be transferred from $\gamma^{(rp-)}$ to $\gamma^{(rp+)}$ in such a way that no two symbols from the same row of $\gamma^{(rp-)}$ go into the same column of $\gamma^{(rp+)}$?" This, however, is just the ordinary T-condition for $\gamma^{(rp-)}$ into $\gamma^{(rp+)}$, and we know that it is satisfied if and only if $\gamma^{(rp+)} \supset \gamma^{(rp-)}$.

The remainder diagrams, however, are not uniquely determined, as the partial transfer of the symbols can in general be carried out in many ways. To satisfy the T-condition, we need only *one* way of doing the partial transfer which will make $\gamma^{(rp+)} \supset \gamma^{(rp-)}$. Since the two parts of the partial transfer are independent of each other, we can transfer the unprimed symbols into $\gamma^{(r+)}$ in such a way as to make $\gamma^{(rp+)}$ as "large" as possible, and similarly transfer the primed symbols so as to make $\gamma^{(rp-)}$ as small as possible. The possibility of doing this depends on the existence of a unique "largest remainder" for the transfer of the unprimed symbols, and a "smallest remainder" for the transfer of the primed symbols. We prove below that such unique largest and smallest remainders exist, and that they are constructed in the following way:

i) Largest Remainder: Given two diagrams $\gamma^{(a)}$ and $\gamma^{(b)}$, with $o_i^{(a)} \geqq o_i^{(b)}$ for all i, the boxes in the first row of $\gamma^{(b)}$ being filled with 1's, the second row with 2's, etc. Transfer the symbols into the boxes of $\gamma^{(a)}$ one at a time, in any order, placing each symbol as far down as possible subject to the condition that it does not go into a column already occupied by a symbol of its own kind. If there are two or more boxes satisfying this condition and equally low down, choose the one farthest to the right. When all the symbols have been transferred, the unfilled boxes will be in diagram form without further rearrangement, and this diagram will be the largest remainder $\gamma_{\max}(\gamma^{(a)}, \gamma^{(b)})$.

The question of the smallest remainder $\gamma^{(rp-)}$ takes a completely analogous form if we reformulate it somewhat. Instead of transferring the symbols from $\gamma^{(p-)}$ into $\gamma^{(r-)}$, we imagine that the boxes of $\gamma^{(r-)}$ are transferred into $\gamma^{(p-)}$, with no two from the same column going into the same row of $\gamma^{(p-)}$. The symbols remaining uncovered by transferred boxes form the remainder. It is obvious that this is equivalent (as far as forming the remainder is concerned) to the formulation in terms of symbol transfer, and that it differs from the situation of (1) above only in the interchange of the roles of rows and columns. The construction of the smallest remainder is therefore carried out as follows:

ii) *Smallest Remainder*: Given two diagrams, $\gamma^{(a)}$ and $\gamma^{(b)}$, with $u_i^{(a)} \geqq u_i^{(b)}$ for all i, and with the first column of $\gamma^{(b)}$ filled with symbols 1, the second with 2's, etc. Transfer the symbols into the boxes of $\gamma^{(a)}$ one at a time, in any order, placing each symbol as far the right as possible consistent with the condition that it not go into a row already occupied by a symbol of its own kind. If there are two or more allowed boxes equally far to the right, choose the one farthest down. When all the symbols have been transferred, the unfilled boxes will be in diagram form without further rearrangement, and this diagram will be the smallest remainder $\gamma_{\min}(\gamma^{(a)}, \gamma^{(b)})$.

Examples of largest and smallest remainder construction are shown in Fig. 6.

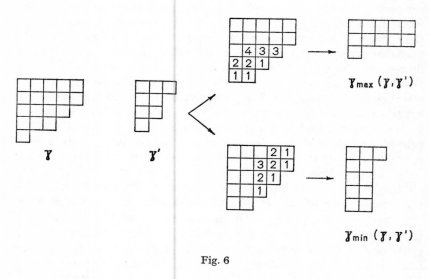

Fig. 6

Proof of the Prescription: In the beginning of this proof we consider the largest remainder prescription with the transfer being carried out in a particular order: First the 1's then the 2's, etc. At the end, we will see that the order is irrelevant.

To prove the validity of the prescription, we start with some definitions and preliminary remarks.

We call two allowed arrangements of the symbols among the boxes "equivalent" if they have the same remainder. Since the remainder, as well as the C-condition, depends only on the number of symbols in each column, we can assume without loss of generality that all symbols in a given column are piled up at the bottom of the column. Now, for fixed number of symbols in each column, we define a "standard" arrangement as one in which the 1's are placed as low down as is compatible with the C-condition and the fixed column distribution (number of symbols in each column), the 2's as low as is compatible with the C-condition, the column distribution, and the placement of the 1's, etc.

Now, starting with an arbitrary arrangement $\mathscr{A}$, we can rearrange it into standard form as follows:

We rearrange the symbols among the occupied boxes so that the 1's are at the bottoms of the $\nu_1^{(b)}$ longest occupied columns (*i.e.*, in the $\nu_1^{(b)}$ lowest-lying occupied boxes with no two in the same column), the 2's are at the bottoms of the $\nu_2^{(b)}$ longest columns of the remainder which still have space available, etc. This standard arrangement $\mathscr{A}_s$ is equivalent to $\mathscr{A}$, as the same boxes are occupied as before, though not necessarily by the same symbols. Now, if we can move an entire pile of symbols from one column into a longer, previously unoccupied one, we obtain an arrangement with a larger remainder than $\mathscr{A}_s$, since we have increased the length of a shorter column of the remainder at the expense of a longer one. Let us do this with all the "1-piles" (piles with 1's at the bottoms), *i.e.*, move the 1-piles into the ν_1 longest columns of $\gamma^{(a)}$, choosing the columns farthest to the right in case of ties. (This will not displace any other piles, as the 1-piles already are in the longest occupied columns.) The resulting arrangement has the 1's placed according to the prescription, and has a larger remainder than does $\mathscr{A}_s$. It is standard as far as the placement of the 1's is concerned; if it is not also standard for the other symbols, we rearrange them as before to form a standard arrangement, $\mathscr{A}'_s$ whose remainder is larger than that of $\mathscr{A}_s$. Similarly, we now move the 2-piles (the 2's, plus whatever else may be on top of them) into longer columns (of $\gamma^{(a)}$ minus the squares already occupied by 1's) where possible and rearrange into a standard arrangement, $\mathscr{A}''_s$, in which the 1's and 2's are arranged according to the prescription and whose remainder is larger than that of $\mathscr{A}'_s$, hence also larger than that of $\mathscr{A}_s$. Continuing this procedure for the 3's, 4's, etc., we end up with the prescribed arrangement. Its remainder is greater than that of $\mathscr{A}_s$, hence also greater than that of the equivalent arrangement $\mathscr{A}$. But $\mathscr{A}$ was arbitrary, so the prescribed arrangement has a larger remainder than any other, and the validity of the prescription is proved.

It is evident from the nature of the proof that the prescription is, as claimed in Section III, independent of the order of placement of the symbols. To prove the prescription for another order of placement, one need only redefine "standard arrangement" in such a way that first preference (*i.e.*, the lowest available places) is given to the symbols which are to go in first, then those which go in next, etc. The validity of the prescription for the smallest remainder now follows by interchanging rows and columns.

From the above discussion, we see immediately the criterion for satisfying the T-condition for $\bar{\gamma}^{(p)}$ into $\bar{\gamma}^{(r)}$. It is satisfied if and only if

$$1) \ \bar{\gamma}^{(r)} \supset \bar{\gamma}^{(p)}$$

and

$$2) \ \gamma_{\max}(\gamma^{(r+)}, \gamma^{(p+)}) \supset \gamma_{\min}(\gamma^{(p-)}, \gamma^{(r-)}).$$

We note that the T-condition in this case is not transitive. For example, referring to Fig. 7, we see that the T-condition is satisfied for $\bar{\gamma}^{(c)}$ into $\bar{\gamma}^{(b)}$, and for $\bar{\gamma}^{(b)}$ into $\bar{\gamma}^{(a)}$, but not for $\bar{\gamma}^{(c)}$ into $\bar{\gamma}^{(a)}$.

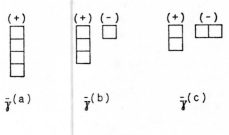

Fig. 7

III. Chirality Functions; Qualitative Completeness

A molecule may be completely specified by describing a skeleton, plus the nature (and perhaps the orientation) of the ligand at each site. Thus, a skeleton may be thought of as defining a class of molecules, with individual members of the class being determined by specification of the ligands on the various sites. A given molecule may belong to more than one such class, depending on which part of the molecule is taken to be the skeleton and which the ligands. For instance, ethane may be thought of as the six-site ethane skeleton with six H-atoms as ligands; or alternatively as the four-site methane skeleton with one methyl and three H-atom ligands.

In the present article, it will be assumed that a molecule within a skeletal class is completely determined for our purposes by specifying the nature of the ligand on each site, though a ligand is permitted to have two enantiomeric (mirror image) forms. If more than one orientation of a ligand on a site is possible, this means that one of the following must hold: either

(a) The ligand must possess sufficient symmetry to make all properties of interest to us invariant under changes of orientation; or

(b) Properties of interest refer to time- or ensemble-averages to which all orientations contribute equally; or

(c) For given distribution of ligands, a single orientation is strongly preferred.

The problem of how to relax this restriction will be alluded to in Section X.

Within a given skeletal class, and subject to the above restriction, any quantitative molecular property will depend on the nature of the ligands on the various sites, and can thus be thought of as a function of the ligands, or of properties (parameters) thereof. A pseudoscalar, or chiral, property is one (such as rotation of the plane of polarized light) whose numerical value is invariant under rotations of the molecule, but which changes sign when the molecule is replaced by its mirror image. A function of the ligands, or of ligand parameters, which describes a pseudoscalar property of molecules within a skeletal class will be called a *chirality function*. A chirality function may, in general, take many forms. It could consist simply of a table of experimental data; or of an empirical function chosen to fit the data; or of the result of a theoretical calculation, etc.

As an example, which will also lead us to the concept of qualitative completeness, consider the allene skeleton, as shown in Fig. 8, and for the moment consider only achiral ligands. Besides the unit element, the symmetry group of the skeleton, D_{2d}, consists of the rotation operations (in permutation group notation) (12)(34), (13)(24), and (14)(23), plus the improper rotations (1)(2)(34), (12)(3)(4), (1324), and (1432).

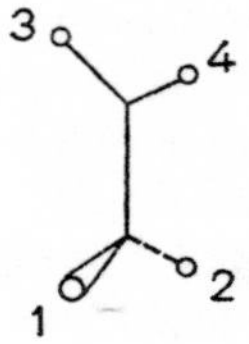

Fig. 8

If λ is some scalar parameter associated with a ligand, then it is readily verified that the functions

$$\chi_1 = (\lambda_1-\lambda_2)(\lambda_3-\lambda_4);$$

$$\chi_2 = (\lambda_1-\lambda_2)(\lambda_1-\lambda_3)(\lambda_1-\lambda_4)(\lambda_2-\lambda_3)(\lambda_2-\lambda_4)(\lambda_3-\lambda_4) \quad (1)$$

are both chirality functions for this skeleton: both χ_1 and χ_2 are unchanged under the rotation operations, and change sign under the improper rotations. One could imagine attempting to find an empirical ligand parameter λ so as to fit a given set of experimental data with a function of the form χ_1 or χ_2.

As it turns out, however, neither of these functions can be used to describe a chiral property of a wide class of molecules without encountering a fundamental difficulty. For example, consider an isomer mixture in which the molecules I, II, III of Fig. 9 appear in equal concentrations. χ_1 for this mixture would be, apart from a multiplicative constant, simply

$$\chi_1(I) + \chi_1(II) + \chi_1(III).$$

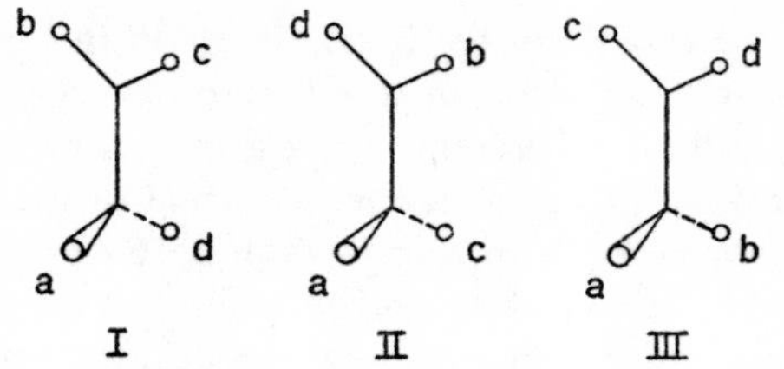

I　　　II　　　III

Fig. 9

But

$$\chi_1(\mathrm{I}) = (\lambda_a-\lambda_d)(\lambda_b-\lambda_c);$$
$$\chi_1(\mathrm{II}) = (\lambda_a-\lambda_c)(\lambda_d-\lambda_b);$$
$$\chi_1(\mathrm{III}) = (\lambda_a-\lambda_b)(\lambda_c-\lambda_d),$$

and it is readily seen that the sum vanishes identically, even though the mixture is not racemic. One varifies easily that χ_2 does not vanish for this mixture. On the other hand, for the chiral molecule of Fig. 10, as well as for an equal mixture of the two isomers I', II' of Fig. 11, one easily shows that χ_2 vanishes identically while χ_1 does not.

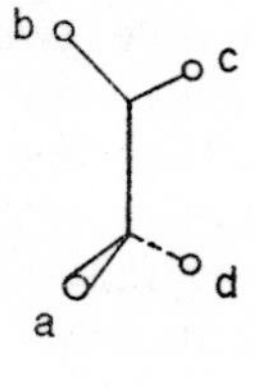

Fig. 10

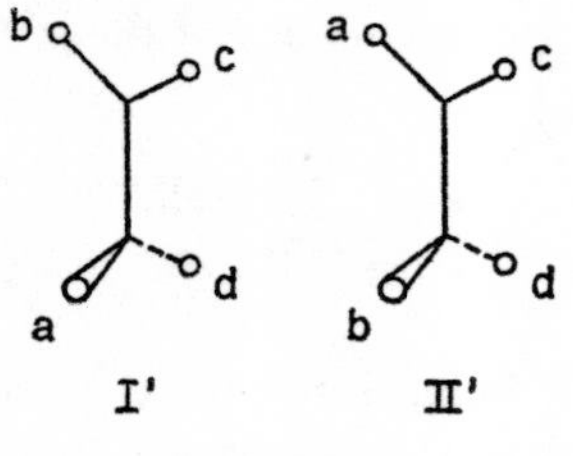

I'　　　II'

Fig. 11

Either of these functions by itself, then, is incapable of giving a sufficiently general description of a chiral property, as each vanishes identically in situations where there is no reason of symmetry why it should, *i.e.*, no reason inherent in the definition of a chirality function. The sum of the two, $\chi_1 + \chi_2$, appears to offer better prospects than either one by itself; but it is not clear at this point whether this function is sufficiently general, or in general, how one decides whether a chirality function will encounter difficulties of this kind or not.

The answer to this problem was given by Ruch and Schönhofer [6] in terms of the concept of "qualitative completeness", which is defined as follows:

A chirality function is said to be qualitatively complete if there is no nonracemic isomer mixture for which it vanishes identically.

To analyze the implications of qualitative completeness, we first make a few preliminary definitions and remarks.

Given a molecule M_0 belonging to a skeleton with n sites, the group which generates all the isomers will be called $\mathfrak{S}$. The group $\mathfrak{S}$ includes all possible rearrangements of the ligands among the sites, as well as all permissable operations (*e.g.*, inversion) which may be performed on a ligand. In this paper, $\mathfrak{S}$ will always be either $\mathfrak{S}_n$ or $\bar{\mathfrak{S}}_n$. $\mathfrak{S}$ possesses a subgroup $\mathfrak{G}$, consisting of those rearrangements which can be interpreted as rotations and/or reflections of the molecule. $\mathfrak{G}$ in turn possesses a subgroup $\mathfrak{D}$, corresponding to pure rotations. $\mathfrak{G}$ is often, but not always, isomorphic with the point group of the skeleton. It fails to be always isomorphic because sometimes two or more elements of the point group may correspond to the same permutation. Thus, an operation of $\mathfrak{D}$ applied to M_0 merely gives the same molecule in a different orientation, while a member of the coset of $\mathfrak{D}$ in $\mathfrak{G}$ gives the mirror image M_0^* in some orientation. A chirality function, by definition, must belong to the "chirality representation" $\Gamma^{(\chi)}$ of $\mathfrak{G}$, which has character $+1$ for members of $\mathfrak{D}$, -1 for members of its coset in $\mathfrak{G}$. The index of $\mathfrak{D}$ in $\mathfrak{G}$ is always 2. If the ligands are allowed to be chiral, it is to be noted that the elements of $\mathfrak{D}$ do not contain τ_0, while those of its coset in $\mathfrak{G}$ do.

A mixture of isomers may be represented by an "ensemble operator", which is defined simply as a member a of the group algebra $\mathfrak{U}$ of $\mathfrak{S}$.

$$a = \sum_{s \in \mathfrak{S}} a(s)s,$$

with the coefficients $a(s)$ being interpreted in terms of concentrations. Applied to a molecule M_0, the operator a generates a set of isomers, sM_0, each formally multiplied by a coefficient $a(s)$. The $a(s)$ are to be interpreted in terms of concentrations, but, since $a(s)$ need not be real and positive for a to be in $\mathfrak{U}$, this requires a little bit of care. There are two ways of interpreting real negative $a(s)$, each equally valid for our purposes, viz:

(i) If $a(s)$ is negative, it is interpreted as a positive concentration $-a(s)$ of the mirror image $(sM_0)^*$. This will obviously produce consistent results for all chirality functions. (This is the interpretation used by Ruch and Schönhofer.[6])

(ii) We can start with a mixture in which all isomers are present in equal concentration, and interpret the $a(s)$ as *concentration increments*, in which case they can obviously be either positive or negative.

Complex $a(s)$ may be interpreted as representing two mixtures simultaneously, one corresponding to the real, one to the imaginary part. This will not cause any inconsistency with the operations we shall be performing with the ensemble operators. Alternatively, we could simply restrict ourselves to real $a(s)$, as the real numbers suffice for the full reduction of representations of $\mathfrak{S}_n$ and $\bar{\mathfrak{S}}_n$.

Given a skeleton with n sites, and a set of ligands (in general all different) $l_1, l_2, \ldots, l_n$, we can form a molecule M_0 by distributing the ligands in an arbitrary way among the sites. A chirality function $\chi(M)$ will have a particular value for this molecule. Alternatively, if we keep the ligands fixed on the same sites but allow parameters characterizing the ligands to vary, we can think of $\chi(M_0)$ as a function of the ligands. The corresponding chirality function for the mixture aM_0 is

$$\chi(aM_0) = \sum_{s \in \mathfrak{S}} a(s)\chi(sM_0) . \qquad (2)$$

Qualitative completeness of χ means that, if a is not the operator for a racemic mixture, then $\chi(aM_0)$, considered as a function of the ligands, does not vanish identically.

To analyze the implications of this, we choose a basis for each irreducible representation $\Gamma^{(r)}$ of $\mathfrak{S}$ such that the first z_r basis functions transform according to $\Gamma^{(\alpha)}$ under $\mathfrak{G}$, where z_r is the number of times $\Gamma^{(\alpha)}$ is subduced by $\Gamma^{(r)}$ in $\mathfrak{G}$. (Of course, some of the z_r may be zero.) Now, since a is a member of $\mathfrak{U}$, it can be expressed in terms of the e-operators as

$$a = \sum_{r,i,j} a_{ij}^{(r)} e_{ij}^{(r)} .$$

That a is nonracemic means that it does not identically annihilate *all* chirality functions, *i.e.*, that at least one of the coefficients $a_{ij}^{(r)}$, with $j \leqslant z_r$, is different from zero. Qualitative completeness for χ means, therefore, that it is not annihilated by any of the $e_{ij}^{(r)}$ with $j \leqslant z_r$, nor by any linear combination of them. This means that χ must possess z_r components $\chi_j^{(r)}$ belonging to each $\Gamma^{(r)}$ of $\mathfrak{S}$. These components must not only be linearly independent, but also all functions $e_{ij}^{(r)} \chi_j^{(r)}$ must be linearly independent. For example, if $z_r = 2$, it will not do to have $\chi_2^{(r)} = e_{21}^{(r)} \chi_1^{(r)}$, since in this case the chirality function $\chi = \chi_1^{(r)} + \chi_2^{(r)}$ would be annihilated by the chiral ensemble operator $a = e_{11}^{(r)} + e_{22}^{(r)} - e_{12}^{(r)} - e_{21}^{(r)}$. This independence can be achieved, for example, by having the different components belonging to the same IR depend on different ligand parameters. Because of this independence, it also follows, according to the reasoning of Section II-C, that the induction from $\mathfrak{G}$ to $\mathfrak{S}$ starting with χ must be regular.

An IR $\Gamma^{(r)}$ of $\mathfrak{S}$ with $z_r \neq 0$ will be called a chiral representation (with respect to a given skeleton).

Our results may be summarized as

Theorem 1: It is necessary and sufficient for qualitative completeness of χ that χ contain z_r independent components transforming according to each IR $\Gamma^{(r)}$ of $\mathfrak{S}$, where z_r is the number of times $\Gamma^{(x)}$ is contained in $\{\Gamma^{(r)}(\mathfrak{S})\}_\mathfrak{G}$. It is also necessary and sufficient for qualitative completeness that the representation of $\mathfrak{S}$ induced by χ is just $[\Gamma^{(x)}(\mathfrak{G})]^\mathfrak{S}$, *i.e.*, that the induction is regular.

For readers unfamiliar with these techniques, it might be helpful at this point to work out an example in some detail. We choose that of the allene skeleton, already discussed somewhat in this section, and at first we limit ourselves to achiral ligands, so that $\mathfrak{S} = \mathfrak{S}_4$. The character table for $\mathfrak{S}_4$ is shown in Table 1. In this case, the subgroup $\mathfrak{G}$ is just D_{2d}, and its rotational subgroup is D_2. Table 2 shows the classes of D_{2d}, the number of elements in each, the class of $\mathfrak{S}_4$ and of $\mathfrak{S}'_4$ to which each belongs, and the character of each for the representation $\Gamma^{(x)}$.

Table 1. Characters of $\mathfrak{S}_4$

Rep. (r)	Young Dgm	c	(1^4)	$(1^2,2)$	$(1,3)$	(2^2)	(4)
		c	1	6	8	3	6
1	(4)	1	1	1	1	1	1
2	(3,1)	3	1	0	-1	-1	-1
3	(2^2)	2	0	-1	0	2	0
4	(2,1)	3	-1	0	-1	-1	1
5	(1^4)	1	-1	1	1	1	-1

Table 2. Some properties of D_{2d}

c	E	C_2	U_2	σ	S_4
c	1	1	2	2	2
$c(\mathfrak{S}_4)$	(1^4)	(2^2)	(2^2)	$(1^2,2)$	(4)
$c(\bar{\mathfrak{S}}_4^1)$	(1^4)	(2^2)	(2^2)	$\tau_0(1^2,2)$	$\tau_0(4)$
$\chi(\Gamma^{(\varkappa)})$	1	1	1	-1	-1

To get the characters for the representation of $\mathfrak{G}$ subduced by a given representation of $\mathfrak{S}$, we just copy down the characters of that representation for the elements of $\mathfrak{S}$ which are also in $\mathfrak{G}$. This is done for the irreducible representations of $\mathfrak{S}_4$ subduced onto D_{2d} in Table 3. Comparing Tables 2 and 3, and using the standard formula for finding the irreducible parts of a representation by means of the characters, we see that the representations subduced by $\Gamma^{(3)}$ and $\Gamma^{(5)}$ contain $\Gamma^{(\varkappa)}$

Table 3. Characters of $\{\Gamma^{(r)}(\mathfrak{S}_4)\}_{D_{2d}}$

r	E	C_2	$2U_2$	2σ	$2S_4$
1	1	1	1	1	1
2	3	-1	-1	1	-1
3	2	2	2	0	0
4	3	-1	-1	-1	1
5	1	1	1	-1	-1

once each, the others not at all. In other words, in this case $z_1 = z_2 = z_4 = 0$, $z_3 = z_5 = 1$. This means that the regular introduction from $\Gamma^{(\varkappa)}$ of D_{2d} to $\mathfrak{S}_4$ gives a representation containing $\Gamma^{(3)}$ and $\Gamma^{(5)}$ once each, and the others not at all. It also means that a qualitatively complete chirality function must have two independent components, one transforming according to $\Gamma^{(3)}$ under $\mathfrak{S}_4$, the other under $\Gamma^{(5)}$.

If we allow the ligands to be chiral, the isomer-generating group $\mathfrak{S}$ becomes $\bar{\mathfrak{S}}_4$. The induction from D_{2d} to $\bar{\mathfrak{S}}_4$ is obtained via $\bar{\mathfrak{S}}_4^1$, according to the procedure outlined in Section II-E. The classes of $\bar{\mathfrak{S}}_4^1$ are those of $\mathfrak{S}_4$, plus each of these multiplied by τ_0. Its characters are those of $\mathfrak{S}_4$ for each class c in $\mathfrak{S}_4$; and the same ones multiplied by ± 1, according as the representation is g or u, for the classes $\tau_0 c$. Otherwise, the procedure is just as before. Fig. 12 shows, in diagramatic form, the representations of $\bar{\mathfrak{S}}_4^1$ induced by $\Gamma^{(\varkappa)}$ of D_{2d}, and the representations of $\bar{\mathfrak{S}}_4$ induced by each of these. In the present case, no representation appears more than once in a single step of the induction. A qualitatively complete chirality function for this case, therefore, must have eleven independent components, one for each two-part diagram appearing on the right-hand side of Fig. 12. Note that this means there must be two components belonging to the two-part diagram $(1^2+;\ 1^2-)$.

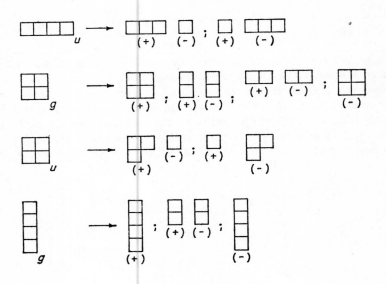

Fig. 12

IV. Simple Explicit Forms of Chirality Functions

As shown in the previous section, a qualitatively complete chirality function contains $\sum_r z_r$ components, so the problem of the explicit construction of χ reduces to that of the construction of its components.

To construct a chirality function belonging to a particular representation $\Gamma^{(r)}$ of $\mathfrak{S}$, one proceeds, in principle, as follows: Starting with an arbitrary function $\psi(1,2,\ldots,n)$, one applies one of the Young operators (arbitrarily chosen) which project onto $\Gamma^{(r)}$. If the result it not zero, it will be a function belonging to $\Gamma^{(r)}$, though not necessarily a chirality function. One then applies $e^{(\chi)}$, the projection operator onto the chiral representation of $\mathfrak{S}$. If the result is still not zero, it will be a chirality function having the desired properties. In mathematical form,

$$\chi^{(r)} = e^{(\chi)} y^{(r)} \psi . \tag{1}$$

If at any point in this procedure one gets zero, of course, it is necessary to start over. With experience, one learns to choose ψ and Y in such a way that this will not happen. If $z_r > 1$, one must construct z_r different functions in this way, independent in the sense discussed in the previous section.

In this process, of course, one is trying to obtain a single function, not a sum of functions for different molecules. This means that the operators s of $\mathfrak{S}$ appearing in Eq. (1) must be interpreted as giving new functions of the same molecule, not the same function of different molecules, as explained in Section II-D. The dependence on the ligands is, of course, the same. In practice, this causes no problems, partly because all the operators involved are sums in which s^{-1} is present along with each s, so one does not have to worry about which interpretation is being used.

In principle, the starting function ψ in Eq. (1) is arbitrary, and an unlimited variety of chirality functions belonging to the same representation is possible. In this section, we wish to consider two particularly simple types of functions, which lend themselves to the fitting of experimental data. We will work out the implications of these procedures for the cases $\mathfrak{S} = \mathfrak{S}_n$ and $\mathfrak{S} = \overline{\mathfrak{S}}_n$.

First Procedure: Polynomial of lowest Possible Order.

A particularly simple type of chirality function is a polynomial in one or more ligand parameters, such as the two functions of Eq. (III-I). Maximum simplicity is obtained if one requires each component of the qualitatively complete χ to be a polynomial of lowest possible order in the parameter(s). We first consider the case $\mathfrak{S} = \mathfrak{S}_n$.

If the ligands are not allowed to be chiral, they can be characterized by a single scalar parameter λ. The starting function ψ in Eq. (1) is chosen to be a homogeneous polynomial in the λ. Since each of the monomial terms in ψ separately yields either zero or a function with the desired property, we can consider ψ to be a monomial without loss of generality. We choose the monomial to be of the lowest order which will not be annihilated by the operations of Eq. (1).

The Young operator Y in (1) antisymmetrizes with respect to permutations of sites in the same column in its tableau. The monomial ψ, therefore, cannot be symmetrical with respect to any two sites in the same column, i.e., it cannot contain the same power of λ for any two such sites. The powers of λ for the sites in a given column, therefore, must all be different. The lowest possible choice consistent with this is that they be 0, 1, 2, ..., μ for a column of length μ. Thus, ψ can be chosen to be independent of λ for sites in the first row of the tableau, and to contain λ for sites in the second row, λ^2 for those in the third, etc. The total order is therefore

$$g = \sum_j (j-1) \nu_j. \tag{2}$$

This is obviously necessary for the existence of a nonvanishing chirality function of the given order; in fact, it is also sufficient. For, if $Y\psi$ does not vanish, it belongs to the representation $\Gamma^{(r)}$, and it follows that a complete basis for $\Gamma^{(r)}$, including the z_r chirality functions, can be generated from it by applying the operations of $\mathfrak{S}_n$, which do not, of course, affect the order of the polynomial. There is, of course, no reason *a priori* why the different components must depend on the same parameter. For different components belonging to the same r, different parameters must be chosen in order for them to have the requisite independence, as discussed in the preceeding section.

It is easy to show that chirality functions obtained according to this first procedure depend only on differences of the λ's. To show this, we make the change of variables

$$y_1 = \lambda_1 + \lambda_2 + \ldots + \lambda_n;$$

$$y_2 = \lambda_2 - \lambda_1;$$

$$y_3 = \lambda_3 - \lambda_2;$$
$$\ldots$$
$$y_n = \lambda_n - \lambda_{n-1}.$$

Our polynomial can be expressed in terms of the y's as well as in the λ's, and is of the same order in both. If any term contains a power of y_1, all the other terms will contain the same power of it, since y_1 is totally symmetric. Also because of this total symmetry, the function obtained by dividing out this power of y_1 will have all the same transformation properties as the original one. The polynomial of lowest order with the given transformation properties, therefore, will not contain y_1 but will be expressible entirely in terms of the differences $y_2, y_3, \ldots, y_n$, which is the result to be proved.

As an example, we work this out for the allene skeleton considered in Section III. According to the results obtained there, a qualitatively complete χ must contain two components, one belonging to $\Gamma^{(3)}$ of Table 1, the other to $\Gamma^{(5)}$. For the representation $\Gamma^{(3)}$, Eq. (2) tells us that $g = 2$. We choose $\psi = \lambda_2 \lambda_4$ and the tableau of Fig. 13. The Young operator (taken to be QP instead of PQ for convenience) is

$$Y = [1 - (12)][1 - (34)][1 + (13)][1 + (24)]. \tag{3}$$

$$\begin{array}{|c|c|} \hline 1 & 3 \\ \hline 2 & 4 \\ \hline \end{array}$$

Fig. 13

Applying this to our ψ, we find

$$Y \lambda_2 \lambda_4 = 4(\lambda_2 - \lambda_1)(\lambda_4 - \lambda_3). \tag{4}$$

For $e^{(\chi)}$ we have[d]

$$e^{(\chi)} = 1 + (12)(34) + (13)(24) + (14)(23)$$
$$- [(34) + (12) + (1324) + (1423)]. \tag{5}$$

---

[d] The projection operator $e^{(\chi)}$ of Eq. (5) is not normalized. Since normalization plays no particular role in the theory, we shall ignore it in the remainder of this section, with the result that some equations are true only up to a multiplicative constant. Such a constant does not affect transformation properties, so the chirality functions obtained retain their validity.

It is readily verified that $e^{(x)}$ applied to $Y\lambda_2\lambda_4$ merely multiplies it by a constant. Dropping a multiplicative constant, then, we can take our function to be

$$\chi^{(3)} = (\lambda_1 - \lambda_2)(\lambda_3 - \lambda_4), \qquad (6)$$

which is identical with χ_1 of Eq. (III-1). Proceeding analogously, one finds that the component belonging to $\Gamma^{(5)}$ by this procedure is just χ_2 of (III-1).

If the ligands are allowed to be chiral, we have $\mathfrak{S} = \bar{\mathfrak{S}}_n$. In this case, a single scalar parameter is not sufficient to characterize a ligand, since that could not distinguish between the ligand and its mirror image. In addition to our scalar parameter λ, therefore, we need a pseudoscalar one, $\varkappa$. In this case, one must use some care in defining what one means by "lowest order", since it may be possible to obtain a decrease in the order of one of the parameters at the expense of an increase in the other. To resolve this question, we note that any even power of $\varkappa$ is a scalar, in other words, a "λ-like" parameter. We can thus define our parameters in such a way that no single $\varkappa$ appears raised to a power higher than the first, by the simple expedient of considering any even power of $\varkappa$ to be a function of λ. This leads us to formulate the problem as follows: The starting function ψ in Eq. (1) is to be chosen to be a monomial in the λ and $\varkappa$, with individual $\varkappa$'s not raised to any power higher than the first. The number of factors $\varkappa$ in ψ is chosen to be as low as possible; then, keeping the $\varkappa$-dependence unchanged, the order in the λ is minimized.

The Young operator $\bar{Y}$ antisymmetrizes with respect to reflections on all sites in the $(-)$ part of the tableaux. This will annihilate any function depending only on the scalar λ for these sites, so there must be a factor of $\varkappa$ in ψ for each of these. The order of $\varkappa$ in ψ is therefore

$$g(\varkappa) = n_-. \qquad (7)$$

The lowest order in the λ is then obtained exactly as in the case of achiral ligands. The result is

$$g(\lambda) = \sum_j (j-1)(\nu_j^{(+)} + \nu_j^{(-)}). \qquad (8)$$

Again, we give an example involving the allene skeleton, this time finding the two components belonging to the representation $(1^2+;\ 1^2-)$ which are present according to Fig. 12. Although it is not necessary, it is convenient to choose these so that they also belong to definite IR's of $\mathfrak{S}_4^!$, one to $(2^2)_g$, the other to $(1^4)_g$. This can be achieved by applying a Young operator to project onto the desired representation of $\mathfrak{S}_4^!$, after having projected onto the representation of $\bar{\mathfrak{S}}_4$, but before projecting onto $\Gamma^{(\chi)}$.

To get the component belonging to $(2^2)_g$ we use $\psi = \varkappa_3\varkappa_4\lambda_2\lambda_4$ and Young operators for the tableaux of Fig. 14. The Young operator for Fig. 14(a) is

$$\bar{Y} = [1 - (12)][1 - (34)](1 + \tau_1)(1 + \tau_2)(1 - \tau_3)(1 - \tau_4) .$$

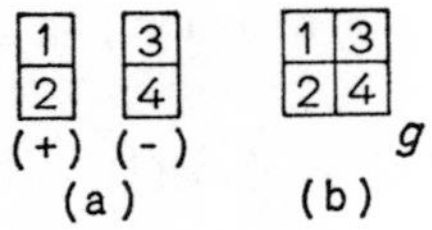

```
┌─┐┌─┐   ┌─┬─┐
│1││3│   │1│3│
├─┤├─┤   ├─┼─┤
│2││4│   │2│4│
└─┘└─┘   └─┴─┘
(+) (-)      g
  (a)      (b)
```

Fig. 14

Applying this to our ψ, we obtain, apart from a multiplicative constant,

$$\bar{Y}\psi = \varkappa_3\varkappa_4(\lambda_2 - \lambda_1)(\lambda_4 - \lambda_3) . \tag{9}$$

The Young operator for Fig. 14(b) is

$$Y = [1 + (13)][1 + (24)][1 - (12)][1 - (34)](1 + \tau_0) .$$

Applied to the function (9), it gives (again apart from a multiplicative constant)

$$Y\bar{Y}\psi = (\varkappa_1\varkappa_2 + \varkappa_3\varkappa_4)(\lambda_1 - \lambda_2)(\lambda_3 - \lambda_4)$$

$$- (\varkappa_1\varkappa_4 + \varkappa_2\varkappa_3)(\lambda_1 - \lambda_4)(\lambda_2 - \lambda_3) \tag{10}$$

Eq. (10) gives us a function belonging to the desired representations of $\bar{\mathfrak{S}}_4$ and of $\mathfrak{S}_4^!$, but it is not yet a chirality function. To get a chirality function, we must apply $e^{(\chi)}$. In this case, we have

$$e^{(\chi)} = 1 + (12)(34) + (13)(24) + (14)(23)$$

$$- \tau_0[(34) + (12) + (1324) + (1423)] . \tag{11}$$

By applying the operator (11) to the function (10) and again dropping a multiplicative constant, we obtain the final result for this component:

$$\chi_a = 2(\varkappa_1\varkappa_2 + \varkappa_3\varkappa_4)(\lambda_1 - \lambda_2)(\lambda_3 - \lambda_4)$$
$$- (\varkappa_1\varkappa_3 + \varkappa_2\varkappa_4)(\lambda_3 - \lambda_1)(\lambda_2 - \lambda_4)$$
$$- (\varkappa_1\varkappa_4 + \varkappa_2\varkappa_3)(\lambda_1 - \lambda_4)(\lambda_2 - \lambda_3) . \quad (12)$$

Proceeding analogously, but using the tableaux of Figs. 14(a) and 15, we obtain for the component belonging to $(1^4)_g$ of $\mathfrak{S}_4^i$:

$$\chi_b = (\varkappa_1'\varkappa_2' + \varkappa_3'\varkappa_4')(\lambda_1' - \lambda_2')(\lambda_3' - \lambda_4')$$
$$+ (\varkappa_1'\varkappa_3' + \varkappa_2'\varkappa_4')(\lambda_3' - \lambda_1')(\lambda_2' - \lambda_4')$$
$$+ (\varkappa_1'\varkappa_4' + \varkappa_2'\varkappa_3')(\lambda_1' - \lambda_4')(\lambda_2' = \lambda_3') . \quad (13)$$

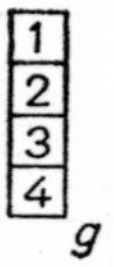

Fig. 15

where the primes on the λ and $\varkappa$ indicate that different variables must be used in the two components in order for them to have the requisite independence.

Second procedure: Function of as few ligands as possible.

In the second procedure, the starting function ψ is not required to be a monomial, but is simply required to depend on as few ligands as possible, being otherwise allowed to be arbitrary. If ψ depends on only h ligands, it must be totally symmetric under both permutations and reflections of the other $(n - h)$. Hence, no two of these $(n - h)$ sites may be in the same column of the tableau of Y, and in the case of chiral ligands, the $(n - h)$ sites must all be in the $(+)$ part of the tableaux. $(n - h)$, therefore, cannot be greater than the number of columns, which is the same as the length of the first row, of the Young diagram (in the case of achiral ligands), or of the $(+)$ part of the Young diagram (in the case of chiral ligands). The lowest value of h is therefore

$$h = n - \nu_1 \qquad \text{(achiral ligands)}; \qquad (14)$$

$$h = n - \nu_1^{(+)} \qquad \text{(chiral ligands)}. \qquad (15)$$

A simple example is the component belonging to $\Gamma^{(3)}$ for the allene skeleton with achiral ligands. We choose $\psi = f(2,4)$, where f is an arbitrary function, and again apply the operators of Eqs. (3), (5). The result is

$$\begin{aligned}\chi^{(3)\prime} &= f(2,4) + f(4,2) - f(1,4) - f(4,1) - f(2,3) - f(3,2) \\ &\quad + f(1,3) + f(3,1) \\ &= g(2,4) - g(1,4) - g(2,3) + g(1,3) \,, \end{aligned} \qquad (16)$$

where $g(j,k) = f(j,k) + f(k,j)$ is a totally symmetric function.

Another version of the second procedure for chiral ligands is to write

$$\psi = \bar{\Psi} \prod_{j \in t\,(-)} x_j$$

and then require $\bar{\Psi}$ to depend on as few ligands as possible. By the same reasoning as before, we find for the lowest possible number of ligands

$$h = n - \nu_1^{(+)} - \nu_1^{(-)} \qquad (17)$$

It is evident that the first procedure is a special case of the second. In the second procedure, the starting function ψ is chosen to depend on as few ligands as possible, but is otherwise arbitrary, while in the first it is further required to be a monomial of lowest possible order in one or two parameters.

The quantity h of the second procedure has a physical interpretation: One must consider interactions between at least h ligands in order to obtain a nonvanishing chirality function belonging to the given representation.

It should be emphasized that these two procedures do not by any means exhaust all possibilities for chirality functions, but are just two particularly simple special cases. In particular, a chirality function obtained by one of these procedures multiplied by an arbitrary function which is totally symmetric under $\mathfrak{S}$ will also be a chirality function having the same transformation properties as the original one, but no longer of lowest order in a parameter or depending on a minimum number of ligands. In particular, if one chiral property is well described by one of these two procedures, others simply related to it may not be. For example, suppose that the molecular rotation at some frequency is described to good approximation by a function obtained by one of these procedures. The specific rotation, which differs from the molecular rotation by a factor of m^{-1}, where m is the mass of the molecule, will then not be a function depending on a minimum number of ligands, containing as it does a factor depending on all the ligands. Only if the masses of all ligands of interest are nearly equal can both properties be expressed in terms of functions obtained by one of these procedures. However, if the description of the phenomenon is only approximate, and if the mass differences do not cause greater errors than are already present, it may be that both properties are describable with about equal accuracy (but with different parameter values) by one of the two procedures.

V. Active and Inactive Ligand Partitions

In this section, we address ourselves to the following question: Suppose we are given a set of n ligands to be placed on our skeleton, some of which are required to be identical (and/or mirror images of one another). Which components, if any, of the qualitatively complete chirality function will vanish identically because of the identity of the ligands? We have already found the answer to this question for the case of the allene skeleton with achiral ligands: If any two ligands are identical, the component $\chi^{(5)}$ vanishes identically, but $\chi^{(3)}$ can be made different from zero with two ligands identical and the other two different, or with two pairs of identical ligands. We now proceed to formulate the question more generally.

An assortment of ligands, some of them identical, can be associated with a Young diagram in a very simple way. For the case of achiral ligands, we define a "ligand partition" as the list of numbers ν_1, ν_2, etc. of identical ligands. Thus, a partition corresponds to a set of ν_1 identical ligands, ν_2 ligands different from the ν_1 previously listed but identical with each other, etc. The sum of the ν's must, of course, be n. The "partition diagram", $\gamma^{(p)}$ is just the Young diagram whose row lengths are ν_1, ν_2, etc.

For the case where the ligands are allowed to be chiral, we proceed analogously. We call two chiral ligands "equivalent" if they are identical up to an inversion, *i.e.*, if they are either identical or mirror images. Now, given an assortment of ligands, some of which are required to be achiral, and some to be identical or equivalent, we define a partition as a list of the numbers ν_1, ν_1, etc., of identical achiral ligands, plus a list of the numbers $\bar{\nu}_1$, $\bar{\nu}_2$, etc., of equivalent chiral ligands. The two-part partition diagram is a diagram $\bar{\gamma}^{(p)}$ whose $(+)$ part has row lengths ν_1, ν_2, etc., and whose $(-)$ part has row lengths $\bar{\nu}_1$, $\bar{\nu}_2$, etc.

Now, given a representation $\Gamma^{(r)}$ of $\mathfrak{S}$ with $z_r \neq 0$, we say that a partition p is "active with respect to $\Gamma^{(r)}$", or "r-active" if there is some molecule belonging to p for which at least one component $\chi_j^{(r)}$ does not vanish identically. A partition is simply called "active" if it is r-active for some r with $z_r \neq 0$. The question posed at the beginning of this section now takes the following form: Given p and r, how do we determine whether p is r-active?

To answer this question, we consider the construction of a chirality function $\chi^{(r)}$ by means of Eq. (IV—1), with the ligand partition p, considering first the case of achiral ligands. For any molecule belonging to p, the starting function ψ will be automatically totally symmetric with respect to permutations of identical ligands. But the Young operator Y antisymmetrizes with respect to permutations of ligands located on sites in the same column of its tableaux. If ψ is not to be annihilated by Y, therefore, it must be possible to distribute the ligands among the sites in such a way that no two identical ligands are on sites in the same column of the tableau of Y. In terms of the partition diagram $\gamma^{(p)}$ this can be rephrased as follows: It must be possible to transfer symbols from $\gamma^{(p)}$ into $\gamma^{(r)}$ in such a way that no two symbols from the same row of $\gamma^{(p)}$ go into the same column of $\gamma^{(r)}$. This, however, is just the transfer condition for $\gamma^{(p)}$ into $\gamma^{(r)}$, as discussed in Section II-F. A necessary condition for r-activity, therefore, is that the transfer condition be satisfied, in other words, according to the results of Section II-F,

$$\gamma^{(r)} \supset \gamma^{(p)}. \tag{1}$$

Eq. (1) is also sufficient for r-activity. For, if (1) is satisfied, for every Young operator $Y^{(r)}$, one can find a molecule belonging to p such that ψ is not annihilated by $Y^{(r)}$. Since the set of all the Young operators projects onto the entire representation space of $\Gamma^{(r)}$, including all the $\chi^{(r)}_j$, it follows that a molecule can be found for which $\chi^{(r)}_j \neq 0$.

For the case of chiral ligands, the answer to the question of r-activity is analogous. For a given partition p, the starting function ψ must be totally symmetric with respect to permutations of identical achiral ligands, and with respect to reflections on sites occupied by achiral ligands. For any pair of equivalent chiral ligands, it must be totally symmetric under transposition of the two (if they are identical), or under transposition followed by reflection of both (if they are mirror images). The Young operator antisymmetrizes with respect to reflections on sites located in the (—) part of its tableau; with respect to permutations among sites in the same column of the tableau; and with respect to transposition between any two sites in the same column followed by reflection on both sites. To have r-activity, therefore, it must be possible to distribute the ligands among the sites in such a way that no two identical or equivalent ligands are put in the same column of the tableau, and that no achiral ligand is on a site in the (—) part of the tableaux. In terms of the partition diagram, this means that symbols must be transferrable from

$\bar{\gamma}^{(p)}$ to $\bar{\gamma}^{(r)}$ with no two symbols from the same row of $\bar{\gamma}^{(p)}$ going into the same column of $\bar{\gamma}^{(r)}$, and no symbol from the $(+)$ part of $\bar{\gamma}^{(p)}$ going into the $(-)$ part of $\bar{\gamma}^{(r)}$. Again, this is just the generalized transfer condition, or

$$\bar{\gamma}^{(r)} \supset \bar{\gamma}^{(p)}. \tag{2}$$

and

$$\gamma_{\max}(\gamma^{(r+)}, \gamma^{(p+)}) \supset \gamma_{\min}(\gamma^{(p-)}, \gamma^{(r-)}). \tag{3}$$

Exactly as in the case of achiral ligands, one shows that (2) and (3) are sufficient as well as necessary for r-activity.

For both situations, therefore, we conclude that a partition p is r-active if and only if the transfer condition is satisfied for the partition diagram of p into the Young diagram of r.

One can also speak of a partition p being active or inactive with respect to an ensemble operator a. p is said to be "a-active" if there is some molecule M belonging to p such that the mixture aM is chiral. Expressing a as before in terms of the e-operators,

$$a = \sum a_{ij}^{(r)} e_{ij}^{(r)},$$

we immediately conclude that a necessary and sufficient condition for a-activity is that there must be at least one $a_{ij}^{(r)} \neq 0$ for $j \leqslant z_r$ and for which p is r-active.

VI. Class-Specific and Ligand-Specific Chirality Functions

For the case of chiral ligands (with which this section is exclusively concerned), it is easy to convince oneself that the representation $(n)_u$ of $\mathfrak{S}_n^1$, whose diagram is shown in Fig. 16, is chiral for every skeleton. For, the representation is totally symmetric under all pure permutations, in particular under those belonging to $\mathfrak{D}$, and antisymmetric under all group elements involving τ_0, in particular under those in the coset of $\mathfrak{D}$ in $\mathfrak{G}$. The representations of $\bar{\mathfrak{S}}_n$ induced by $(n)_u$ are

$$(n)_u \rightarrow [(n-1)+;1-]\,, \ [(n-3)+;3-]\,, \ \text{etc.} \tag{1}$$

Fig. 16

For every skeleton with n sites, therefore, there is at least one component of a qualitatively complete chirality function belonging to each of the representations of Eq. (1). Explicitly, both the first procedure and the alternative version of the second procedure of Section IV give the chirality functions

$$[(n-1)+;1-] : \chi = \sum_j \varkappa_j \,;$$

$$[(n-3)+;3-] : \chi = \sum_{j \neq k \neq l} \varkappa_j \varkappa_k \varkappa_l \,, \ \text{etc.} \tag{2}$$

Since these functions are present for every skeleton, they are characteristic of the ligands, and not of the skeleton or the molecular class determined by it. We call them ligand-specific or $\mathscr{L}$-chirality functions. All other components are characteristic of the skeleton, and we call them class-specific or $\mathscr{C}$-chirality functions.

Two remarks are worth making. First, one easily concludes from the rules derived in the last section that any ligand partition containing at least one chiral ligand is active with respect to the representation $[(n-1)+;1-]$, any partition with at least three chiral ligands is active with respect to $[(n-1)+;1-]$ and $[(n-3)+;3-]$, etc. Thus, $\mathscr{L}$-chirality functions are always present for any skeleton if there are chiral ligands. Second, it should be noted that the representations of $\overline{\mathfrak{S}}_n$ shown in Eq. (1) may, for particular skeletons, also correspond to $\mathscr{C}$-chirality functions. For example, if the representation $(n-1,1)_u$ of $\mathfrak{S}_n^1$ is ciral, it will also induce $[(n-1)+;1-]$ of $\overline{\mathfrak{S}}_n$, and the corresponding chirality function will be class-specific. In general, if $z_r = 1$ for a representation r of $\overline{\mathfrak{S}}_n$ shown in Eq. (1), then the chirality function is ligand-specific. If $z_r > 1$, then there is one ligand-specific component, with the rest being class-specific. We can write

$$z_r = z_r + z_r^{\mathscr{L}},$$

where $z_r^{\mathscr{L}} = 1$ for the representations of Eq. (1), and is zero for the others.

VII. Chirality Numbers

A. Achiral Ligands

For a given skeleton, the set of chiral Young diagrams (representations) permits us to define a set of numbers which are characteristic of the chiral properties of the skeleton. In this subsection, we consider the properties of these numbers for the case of achiral ligands, $\mathfrak{S} = \mathfrak{S}_n$.

For a given Young diagram, we define the numbers o and u as the lengths of the first row and the first column, respectively. For a given skeleton, we define the four "chirality numbers"

$$o_{max}, o_{min}, u_{max}, u_{min},$$

with the maxima and minima being taken over $\gamma^{(r)}$ with $z_r \neq 0$.

From the condition $\gamma^{(r)} \supset \gamma^{(p)}$ for r-activity, it is easy to show that

(a) o_{max} is the largest number of identical ligands which may be present in a chiral molecule. Ruch and Schönhofer [6] have named o_{max} the "chirality order", and abbreviated it simply with o.

(b) The minimum number of different kinds of ligands in a chiral molecule is u_{min}. u_{min}, or simply u for short, is called the "chirality index". [6]

(c) o_{min} is the maximum number of identical ligands which can be present in a partition active with respect to all chiral representations of $\mathfrak{S}_n$.

(d) u_{max} is the minimum number of different kinds of ligands which must be present in a partition active with respect to all chiral representations of $\mathfrak{S}_n$.

Ruch and Schönhofer [6] define a "chiral class" of molecules as one whose skeleton permits of chiral molecules with exclusively achiral ligands. For such classes, they prove the following:

$$n - 3 \leqslant o \leqslant n, \text{ and } 1 \leqslant u \leqslant 4.$$

We give a proof, which is in large part a verbatim translation of that of Ruch and Schönhofer. [6]

For classes with fewer than four sites, the assertion is trivial. For chiral classes with four or more sites, there is at least one triple of sites which does not lie in a symmetry plane of the skeleton. For, if all sites lie in a common symmetry plane, molecules of the class with the ligands all different would possess planes of symmetry, *i.e.*, the class would not be chiral. On the other hand, suppose that the sites do not lie all in a common mirror plane, but that nevertheless every triple of sites lies in a symmetry plane. It follows that every pair of sites lies on the intersection of two different symmetry planes, therefore on an axis of symmetry of the skeleton. But if more than four sites all lie pairwise on an axis of symmetry of a finite figure, they must all lie on a common axis, and the class is again achiral. For chiral classes, then, there is at least one triple of sites which does not lie on a plane of symmetry of the skeleton. Now consider a molecule in which the sites of this triple are occupied by ligands of three different kinds, the other sites by ligands different from these three, but identical with each other. Such a molecule is chiral, since the only improper operation which leaves the three different ligands invariant is a reflection in the plane of the triple, and this changes the rest of the molecule. The assertion follows immediately.

It is evident that the case $o = n$ corresponds to chiral skeletons, which lead to chiral molecules even if all ligands are identical.

B. Chiral Ligands

Also if the ligands are permitted to be chiral, we may easily use the T-condition to determine maximum numbers for identical or equivalent ligands and minimum numbers for types of nonequivalent ligands in molecules with r-active ligand partitions $\bar{\gamma}^{(p)}$.

We need only the relation $\bar{\gamma}^{(r)} \supset \bar{\gamma}^{(p)}$ to derive the following relations for the lengths of first columns and rows.

$$o^{(r+)} \geqslant o^{(p+)},\ u^{(r-)} \leqslant u^{(p-)},\ o^{(r+)} + o^{(r-)} \geqslant o^{(p-)},$$

$$\max(u^{(r+)}, u^{(r-)}) \leqslant u^{(p+)} + u^{(p-)}$$

All these limiting numbers are attained in at least one r-active partition; the first and the second inequalities become equalities for $\bar{\gamma}^{(r)} = \bar{\gamma}^{(p)}$ representing an r-active partition. The third and fourth inequalities become equalities for the diagram $\bar{\gamma}^{(p)}$ whose plus part vanishes and whose minus part consists of the columns of $\gamma^{(r+)}$ and $\gamma^{(r-)}$ properly ordered. It is active as it is smaller than $\bar{\gamma}^{(r)}$ and $\max(\gamma^{(r+)}, \gamma^{(p+)}) =$

$\bar{\gamma}^{(r)} \supset \min(\gamma^{(r-)}, \gamma^{(p-)})$. Therefore we may conclude that for r-active molecules

$o^{(r+)}$ is the maximum number of identical achiral ligands

$o^{(r+)} + o^{(r-)}$ is the maximum number of equivalent chiral ligands

$u^{(r-)}$ is the minimum number of inequivalent chiral ligands

$\max(u^{(r+)}, u^{(r-)})$ is the minimum number of inequivalent ligands.

Each of these numbers occur in at least one r-active ligand partition.

By forming the maxima or minima respectively of these quantities, the extrema being taken over all r with $z_r^\mathscr{C} \neq 0$, we get chirality numbers which are characteristic properties of the skeleton. It should be emphasized that the condition is $z_r^\mathscr{C} \neq 0$ and not $z_r \neq 0$ because the latter would only lead to trivial numbers which express the fact that a ligand partition is active for any achiral frame if it contains at least one chiral ligand. The nontrivial numbers we want should present information about the pseudoscalar properties of the particular molecular class in question. We find these chirality numbers from the following maxima and minima:

chirality order $o^+ = \max \{o^{(r+)}\}$

$\bar{o} = \max \{o^{(r+)} + o^{(-)}\}$

chirality index $u^- = \min \{u^{(r-)}\}$

$\bar{u} = \min \{\max(u^{(r+)}, u^{(r-)})\}$

maxima and minima are to be taken over all r with $z_r^\mathscr{C} \neq 0$.

We call a molecule $\mathscr{C}$-chiral if it possesses a nonvanishing $\mathscr{C}$-chirality function. Correspondingly we call a ligand partition $\mathscr{C}$-active if there are $\mathscr{C}$-chiral molecules belonging to this ligand partition. Now we may formulate the meaning of the chirality numbers as follows:

o^+ is the maximum number of identical achiral ligands which may be present in a $\mathscr{C}$-chiral molecule.

$\bar{o}$ is the maximum number of equivalent ligands which may be present in a $\mathscr{C}$-chiral molecule.

u^- is the minimum number of inequivalent chiral ligands which must be present in a $\mathscr{C}$-chiral molecule.

$\bar{u}$ is the minimum number of inequivalent ligands which must be present in a $\mathscr{C}$-chiral molecule.

In a chiral class, which by definition contains chiral molecules with exclusively achiral ligands, all these chiral molecules are $\mathscr{C}$-chiral. In an achiral class the chirality is only due to $\mathscr{L}$-chirality. Therefore, we have for any class

$$u^- = 0$$

For chiral classes we showed above, under the restriction to achiral ligands, that the chirality order has the lower limit $n-3$ and the upper limit n which characterizes chiral skeletons. If chiral ligands are admitted we have the relation $\bar{o} \geqslant o^+$ and therefore the same limits for $\bar{o}$.

$$n - 3 \leqslant \bar{o} \leqslant n.$$

VIII. Homochirality

In order to understand some of the applications of the next section, it is necessary to discuss a distinction between different skeletons, originally due to Ruch.[16] In this section, ligands are assumed to be achiral.

The question posed and answered by Ruch is the following: Given a molecular class determined by a skeleton, when is it possible to divide all chiral molecules of the class unambiguously and self-consistently into two subclasses, which can be designated as right and left-handed? Ruch is fond of comparing this distinction between skeletons with that between shoes and potatoes. Although different shoes may differ considerably in size, shape, color, etc., one has no difficulty in deciding whether a given shoe is a right or a left one, even if its mate is not available for comparison. For potatoes, however, no such distinction is possible without arbitrariness. The question, therefore, is: which skeletons are "shoe-like", in the sense of permitting a satisfactory division of chiral molecules into "right" and "left" ones, and which are "potato-like", permitting no such distinction?

To answer the question, we must first consider more precisely what is required of a "satisfactory" division of molecules of a class into right and left. We begin by noting that it is always possible in principle to specify an achiral ligand completely by means of a single scalar parameter λ. For example, one could simply list all the possible ligands in some order (*e.g.*, alphabetical), and arbitrarily assign to them the values $\lambda = 1, 2, 3$, etc.[e] By means of this parametrization, we can think of a molecule as being determined by specifying the value of λ at each of the n sites, *i.e.*, as corresponding to a point in an n-dimensional λ-space. By continuously varying the λ's, we can (mentally) transform any molecule of the class continuously into any other one without leaving the class.

---

[e] The question of whether properties of interest are described by any *simple* functions of the λ, such as polynomials generated by the first procedure of Section IV, is, of course, a quite different one, but irrelevant for the purposes of this section.

An acceptable division of chiral molecules into right and left means a division of the λ-space into two regions (say R and L), such that (i) every chiral molecule is in either R or L, and not on the boundary between them; (ii) if a given chiral molecule is in R, then its mirror image is in L, and vice versa; and (iii) achiral molecules are in neither R or L, but on the boundary between them.[f)]

According to the above criteria, the boundary between the regions R and L must be just the subspace of the achiral molecules. On the other hand, the boundary between two regions of the λ-space must necessarily be $(n-1)$-dimensional. Thus, the requirements can only be satisfied if the subset of the achiral molecules is a $(n-1)$-dimensional hypersurface, or a set of such surfaces.

By definition, a molecule is achiral if it is left invariant by some improper operation (reflection or rotary reflection) of the point group of the skeleton. Writing the permutation s corresponding to a given improper operation in cyclic form,

$$s = (1,2,\ldots,f)(f+1,f+2,\ldots,f+g)(\ldots)\ldots, \qquad (1)$$

one sees immediately that a molecule will be left invariant by s if and only if sites in the same cycle are occupied by identical ligands, *i.e.*, if and only if

$$\lambda_1 = \lambda_2 = \lambda_3 = \ldots = \lambda_f ;$$
$$\lambda_{f+1} = \lambda_{f+2} = \ldots = \lambda_{f+g} ; \text{ etc.} \qquad (2)$$

If the permutation s consists of h cycles (including 1-cycles), the subspace in which Eq. (2) is satisfied is h-dimensional. The dimension h is equal to $(n-1)$ if and only if s consists of a single 2-cycle and $(n-2)$ 1-cycles, *i.e.*, if and only if s is a transposition.

Now let $\mathscr{K}$ denote the set of all pairs (ij) of sites such that the transposition (ij) corresponds to an improper operation. The set of $(n-1)$-dimensional hypersurfaces determined by

$$\lambda_j = \lambda_k \qquad (3)$$

for each pair $(jk) \in \mathscr{K}$, are subspaces corresponding to achiral molecules. If the hypersurfaces determined by (3) contain *all* achiral molecules, then the subset of the achiral molecules will indeed be a set of $(n-1)$-

---

[f)] It is clear that, for a chiral class, the subset of the λ-space corresponding to chiral molecules is n-dimensional, while the subset of achiral molecules, which require the equality of two or more λ's, will be of dimension less than n.

dimensional hypersurfaces. This will be true if and only if the subspaces determined by Eq. (2) are all subspaces of those determined by (3), *i.e.*, if the following is satisfied:

(A) Every cycle of every permutation s corresponding to an improper operation of the skeleton point group must contain at least two sites, j,k such that $(jk) \in \mathscr{K}$.

An equivalent way of stating the criterion is

(A') Every achiral molecule of the class must have at least one symmetry operation which in permutational form corresponds to a transposition.

If (A) is satisfied, then we can choose the surfaces (3) as our boundary, designate one region between them arbitrarily as R, and then determine the rest of the space by requiring that one always changes from R to L, or vice versa, when one crosses one of the surfaces (3). This evidently gives a division of the λ-space into regions satisfying criteria (i) and (iii) above. However, it is easy to see that (ii) is also satisfied. For, if $(ij) \in \mathscr{K}$, then mirror image molecules correspond to interchanged values of λ_j and λ_k, and the corresponding points will always lie on opposite sides of the surface (3).

According to our criterion, therefore, we can divide all skeletons into two categories, viz:

(a) "Shoe-like" skeletons, for which condition (A) is satisfied, and an acceptable classification into R and L molecules is possible.

(b) "Potato-like" skeletons, for which condition (A) fails to hold, and no satisfactory classification into R and L is possible.

It should be emphasized that, for skeletons of category (a), the classification into R and L is not necessarily unique. By defining the parameter λ differently, for example, one may arrive at different classifications, all satisfying requirements (i—iii) above.

For skeletons in category (a), with respect to a particular classification, molecules belonging to the same region (R or L) of the λ-space are referred to as "homochiral", those belonging to different regions as "heterochiral". Thus, all right shoes are homochiral with one another.

An example of a skeleton of category (a) is the allene skeletons of Fig. 7. For this skeleton, the surfaces determined by the improper operations (12) and (34) are

$$\lambda_1 = \lambda_2 \text{ and}$$

$$\lambda_3 = \lambda_4 .$$

The operations (1324) and (2423) both determine the one-dimensional space

$$\lambda_1 = \lambda_2 = \lambda_3 = \lambda_4 ,$$

which is a subspace of the above.

An example of category (b) is the four-site skeleton of Fig. 17, with symmetry C_{4v}. Here the improper operations (24) and (13) determine the $(n-1) = 3$-dimensional hypersurfaces

$$\lambda_1 = \lambda_3 \quad \text{and} \quad \lambda_2 = \lambda_4 \,.$$

Fig. 17

The operations (12)(34) and (14)(23), on the other hand, determine 2-dimensional hypersurfaces

$$\lambda_1 = \lambda_2, \ \lambda_4 = \lambda_3 \text{ and}$$
$$\lambda_1 = \lambda_4, \ \lambda_2 = \lambda_3,$$

which are not subspaces of the above.

For skeletons of category (b), certain chiral molecules may be transformed into their mirror images without passing through an achiral state. For, consider an achiral subspace $\mathfrak{U}$ of dimension less than $(n-1)$. By varying the λ's slightly from their values in $\mathfrak{U}$, we can create mirror-image molecules M and M^* in the neighborhood of $\mathfrak{U}$. Now, since $\mathfrak{U}$ is less than $(n-1)$-dimensional, we can transform M into M^* without passing through any point of $\mathfrak{U}$, which is just the assertion to be proved. Since any chirality function must change sign in passing from M to M^*, this means that it must pass through zero at some point on the path between the two. Since all points on the path represent chiral molecules, it follows that the chirality function must vanish for some chiral molecule. For skeletons of category (b), therefore, all chirality functions possess "chiral zeroes", *i.e.*, all vanish for some chiral molecules.

It should be emphasized that the existence of chiral zeroes is different from the vanishing of a chirality function due to a lack of qualitative completeness. When a function is not qualitatively complete, it will vanish *identically* for a wide class of chiral molecules and/or mixtures. A chiral zero of the type mentioned here, however, depends on particular values of the parameters, and one can get away from the zero by varying the parameters slightly, without changing any symmetry property.

IX. Some Experimental and Theoretical Applications

As the theory discussed in this article is still relatively new, the applications made of it to date have been limited, and have thus far been confined to the case of achiral ligands. Haase and Ruch have given quantum mechanical treatments of the methane[17] and allene[18] skeletons. Experimental measurements involving the same two skeletons have been published by Richter, Richter, and Ruch [19], and by Ruch, Runge, and Kresze [20], respectively.

For the methane skeleton (Fig. 18), there is just one chiral representation of $\mathfrak{S}_4$, namely the totally antisymmetric representation (1^4), number

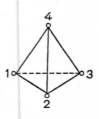

Fig. 18

5 in Table 1. The chirality functions obtained by the first and second procedures, respectively, of Section IV, are:

$$\chi_1^{(5)} = (\mu_1 - \mu_2)(\mu_1 - \mu_3)(\mu_1 - \mu_4)(\mu_2 - \mu_3)(\mu_2 - \mu_4)(\mu_3 - \mu_4); \quad (1)$$

$$\chi_2^{(5)} = \phi(1,2,3) - \phi(2,3,4) + \phi(3,4,1) - \phi(4,1,2), \quad (2)$$

where μ is a scalar parameter specifying the ligands, and $\phi(x,y,z)$ is a function of three ligands totally antisymmetric under permutations of its arguments.

The allene skeleton has, as shown in Section III, two chiral representations: (1^4), the same one which is chiral for methane, and (2^2), number 3 in Table 1. The former of these leads to the same functions

(1) and (2) when the procedures of Section IV are applied (perhaps with different interpretation of the parameter μ and/or of the function ϕ); as already shown in Section IV, application of the two procedures for representation 3 yields:

$$\chi_1^{(3)} = (\lambda_1 - \lambda_2)(\lambda_3 - \lambda_4); \tag{3}$$

$$\chi_2^{(3)} = g(2,4) - g(1,4) - g(2,3) + g(1,3), \tag{4}$$

where λ is a scalar parameter (in general $\neq \mu$), and $g(a,b) = g(b,a)$.

The functions defined by Eqs. (1) and (2), as well as linear combinations of them, satisfy the easily verifiable identity

$$\chi(a,b,c,d) + \chi(b,x,c,d) + \chi(x,a,c,d) + \chi(a,x,b,d) + \chi(a,b,x,c) = 0, \tag{5}$$

where (a,b,c,d) denotes a molecule with the sites 1,2,3,4 occupied by ligands a,b,c,d respectively. The identity (5) holds for the functions (1), (2) for any set of five different ligands a,b,c,d,x. Similarly, the functions (3) and (4) satisfy

$$\chi(a,b,c,d) + \chi(a,c,d,b) + \chi(a,d,b,c) = 0, \tag{6}$$

as already remarked in Section III, as well as

$$\chi(a,b,c,d) = \chi(x,b,y,d) + \chi(x,b,c,y) + \chi(a,x,y,d) + \chi(a,x,c,y). \tag{7}$$

In their treatment of the methane derivatives, Haase and Ruch [17] began by rearranging their Hamiltonian in such a way that it could be expressed as a zero-order hamiltonian representing noninteracting ligands plus a perturbation corresponding to interaction between ligands. They obtained an expression for the optical rotation (for frequencies far from the anomolous dispersion region) using the Rosenfeld equation and perturbation theory through the second order. The problem was set up in such a way that the positions of the ligands were not required to be related to one another by operations of T_d, *i.e.*, it was not ruled out that the skeleton might be distorted from T_d-symmetry. If T_d-symmetry does hold, the theory of Section IV tells us that interaction between at least three different ligands is required to obtain a nonvanishing chirality function, *i.e.*, that one must go at least to the second order of perturbation theory. Thus, terms that appear in first order are to be interpreted as representing deviations from T_d-symmetry. The second-order result could be readily expressed in the form (2), but could be put into the form (1) only by making further, rather drastic, simplifications. One concludes, therefore, that, if higher-order contributions are assumed small, the observed rotation should consist of two components: a "T_d-component" satisfying (5), and another component representing the deviation from T_d symmetry.

The quantity measured in the experimental work on the methane derivatives was the rotation of the *Na D*-line in ethanol solution (sometimes it was necessary to use another solvent, in which case a correction was applied). The sum (5), as well as its separate terms, was evaluated for 13 different choices of the set of ligands a,b,c,d,x. For eleven of these, the observed sum was less in absolute value than its statistical average calculated from the absolute values of the separate terms. For the other two (as well as for some of the eleven), the mixture contained molecules for which one would expect large deviations from T_d-symmetry, and/or dimerization. For those mixtures for which the sum (5) was small, a least-square fit was made to the data with a function of the form (2). This best fit was interpreted as the T_d-component, the remainder as the result of deviation from T_d-symmetry for each molecule. A fit was also made with functions of the form (1), with less quantitative success.

In the case of the allene derivatives, one expects to obtain a function of the form (4) in first-order perturbation theory. Since other terms, including those of the form (2), are of higher order, one might expect this term to be dominant. The quantum-mechanical calculation [18] does indeed yield a term of form (4) in first order. Moreover, it is possible without further drastic approximation to put the first-order result in the form (3), with the parameter λ being expressible in terms of the transverse and axial polarizabilities of the ligand, and various geometrical parameters of the skeleton. Accordingly, in the experimental work [20], an attempt was made to fit the data with a function of the form (3). The ligands used were H, C_6H_6, COOH, CH_3, and C_2H_5. λ-values were measured empirically by measuring the *D*-line rotation for four isomers having at least two ligands equal, and for which therefore the (1^4) component of the qualitatively complete chirality function vanishes identically. Since χ depends only on differences in the λ's, $\lambda(H)$ was arbitrarily set $= 0$. These λ's were then used to calculate the function (3) for four other isomers with these ligands (each with the ligands all different), and the results compared with experiment. For the molar rotation, the largest percent deviation observed was 9.1%, the next largest 3.6%, with the other two being less than 1%. The same procedure was carried out for the specific rotation (leading, of course, to different values for the λ's). In this case, the two largest deviations were 4.5% and 3.9%, with the others again less than 1%. Of course, the legitimacy of using the form (3) as an approximation for both molar and specific rotation is limited by the remark made at the end of Section IV. In any case, the agreement, though admittedly for only a limited class of compounds, is surprisingly good.

Both the allene and the methane skeletons belong to category (a) of Section VIII. Hence, a chirality function can be used to divide the corresponding class of molecules into R and L species according to the sign of the chirality function, with the function vanishing for achiral molecules only. For the molecules experimentally studied in references [19] and [20], the D-line rotation of the allene derivatives is dominated by a term of the form (3), while that of the methane derivatives is at least roughly fitted by one of the form (1). Both of these functions have the property of vanishing only for achiral molecules of the given skeleton, so they are possible candidates for functions to be used for defining R and L species. Accordingly, the authors suggest that, for the limited class of molecules considered, a rational division into R and L molecules is possible on the basis of the sign of the D-line rotation, as long as it is not too small. It should be noted, of course, that the sign of the rotation at a particular frequency can never provide a classification into R and L for all possible molecules belonging to a given skeleton. Whatever frequency one chooses, one can always imagine a hypothetical chiral molecule for which the chosen frequency is in the anomalous dispersion region; by varying the parameters of such a molecule continuously, one can cause the rotation at the chosen frequency to pass through zero for a chiral molecule.

X. Unsolved Problems; Discussion

In the case of $\mathfrak{S} = \tilde{\mathfrak{S}}_n$, where the ligands are allowed to be chiral, we defined a ligand partition in terms of the number of equivalent ligands, rather than the number of identical ones, so that conversion of one or more ligands into their mirror images does not change the partition. With this definition of partition, the question of r-activity could be answered in terms of the T-condition. The question could, however, equally well be posed in terms of identical ligands, so that a partition included the specification of which equivalent chiral ligands were identical and which mirror images. As of this writing, no general solution has been given to the problem of which partitions (in this sense) are active with respect to a given chiral representation, and which inactive. The most important special case is that in which the chiral ligands are required to be pairwise mirror images, so that the ligand assortment is racemic. For such a racemic ligand assortment, it is easy to show [7] that the $\mathscr{L}$-chirality functions vanish identically. But it is not yet known in general which $\mathscr{C}$-chirality functions vanish identically for racemic ligand assortments, and which do not.

It is evident that methods analogous to the ones developed here could be applied to molecular properties which, instead of being pseudoscalar, belong to some other representation of the skeleton point group (vector, tensor, etc. properties). To treat such properties, one needs only to induce from a different representation of $\mathfrak{G}$ than the chiral one.

It has been assumed throughout in this article that a molecule is completely specified when it is specified which ligand (and which enantiomeric form of a chiral ligands) is located on each site. The theory cannot be applied directly, for example, if there are several possible orientations for a ligand on a given site. In this more general case, the group $\mathfrak{S}$ generating all the isomers must include not only permutations and reflections of ligands, but also changes in their orientations leading to different conformations of the molecule. For at least some cases of this type. $\mathfrak{S}$ is a group of a general type known as wreath product groups. [21,22] The representation theory of these groups has been studied, and can be formulated in terms of generalized Young diagrams. Thus, it may be possible to extend the theory to cases of this type, but this program has not as yet been carried through.

Acknowledgement. The author is extremely grateful to Professor Ernst Ruch and the staff of the Institut für Quantenchemie der Freien Universität Berlin for an exceedingly enjoyable and educational sabbatical year in 1971—72, as well as for their continuing friendship and helpful cooperation.

XI. Appendix: Results for Some Special Skeletons

A. Skeletons Considered

3-Site skeleton:
i) Fig. 19, symmetry C_{3v}.

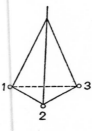

Fig. 19

4-Site skeletons:
i) Fig. 20, triangular pyramid with three equivalent sites and the fourth inequivalent to these. This is an example of a skeleton with a $\mathscr{C}$-chirality function transforming according to $(3+; 1-)$.

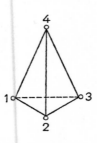

Fig. 20

iii) Allene skeleton, Fig. 7, symmetry D_{2d}.
iv) Fig. 17, C_{4v} symmetry.
v) Fig. 18, T_d symmetry.
6-site skeletons:

vi) Freely rotating ethane, Fig. 21. The symmetry group consists of products of elements from D_{3h} or D_{3d} with internal rotations, and is of order 36. The chiral representation is the usual chiral one under D_{3h} or D_{3d} and has character $+1$ for the internal rotations. One obtains the same homomorphic image $\mathfrak{G}$ whether one uses D_{3h} or D_{3d}.

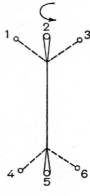

Fig. 21

vii) Fig. 22, symmetry O_h.

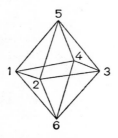

Fig. 22

viii) Cyclopropane derivatives, symmetry D_{3h}, Fig. 23.

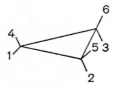

Fig. 23

B. Chirality Functions

Table 4 lists, for skeletons i—v, the $\mathscr{C}$-chirality functions obtained by the first procedure (χ_1) and the alternative version of the second procedure (χ_2) of Section IV, together with the representation of $\mathfrak{S}_n^!$ and of $\overline{\mathfrak{S}}_n$ to which each belongs. χ_2 is omitted in cases where $h=0$ or 1, since in this case it has the same form as χ_1. If the ligands are required to be achiral, only the functions belonging to representations with no $(-)$ part in their Young diagrams remain. The parameters λ are scalar parameters, the $\varkappa$ are pseudoscalar. Different parameters may be used for different representations.

Table 4. Chirality functions

Skeleton	Rep. of $\mathfrak{S}_n^!$	Rep. of $\overline{\mathfrak{S}}_n$	$\mathscr{C}$-Chirality functions
(i)	$(1^3)_g$	(1^3+)	$\chi_1 = (\lambda_1-\lambda_2)(\lambda_2-\lambda_3)(\lambda_3-\lambda_1)$ $\chi_2 = \omega(1,2) + \omega(2,3) + \omega(1,3)$ with $\omega(j,k) = -\omega(k,j)$
		$(1+; 1^2-)$	$\chi_1 = \varkappa_1\varkappa_2(\lambda_1-\lambda_2) + \varkappa_2\varkappa_3(\lambda_2-\lambda_3)$ $+ \varkappa_3\varkappa_1(\lambda_3-\lambda_1)$
(ii)	$(3,1)_u$	$(3+; 1-)$	$\chi_1 = \varkappa_1 + \varkappa_2 + \varkappa_3 - 3\varkappa_4$
		$(2,1+; 1-)$	$\chi_1 = \varkappa_1(2\lambda_4-\lambda_2-\lambda_3) + \varkappa_2(2\lambda_4-\lambda_1-\lambda_3)$ $+ \varkappa_3(2\lambda_4-\lambda_1-\lambda_2)$
		$(1+; 3-)$	$\chi_1 = \varkappa_1\varkappa_2\varkappa_4 + \varkappa_1\varkappa_3\varkappa_4 + \varkappa_2\varkappa_3\varkappa_4$ $- 3\varkappa_1\varkappa_2\varkappa_3$
		$(1+; 2,1-)$	$\chi_1 = \varkappa_4[\varkappa_1\varkappa_2(2\lambda_4-\lambda_1-\lambda_2)$ $+ \varkappa_2\varkappa_3(2\lambda_4-\lambda_2-\lambda_3) + \varkappa_1\varkappa_3(2\lambda_4-\lambda_1-\lambda_3)]$
	$(2,1^2)_g$	$(2,1^2+)$	$\chi_1 = (\lambda_1-\lambda_2)(\lambda_2-\lambda_3)(\lambda_3-\lambda_1)$ $\chi_2 = \omega(1,2) + \omega(2,3) + \omega(3,1)$ with $\omega(j,k) = -\omega(k,j)$
		$(2+; 1^2-)$	$\chi_1 = \varkappa_1\varkappa_2(\lambda_1-\lambda_2) + \varkappa_2\varkappa_3(\lambda_2-\lambda_3)$ $+ \varkappa_3\varkappa_1(\lambda_3-\lambda_1)$
		$(1^2+; 2-)$	$\chi_1 = \varkappa_1\varkappa_4(\lambda_2-\lambda_3) + \varkappa_2\varkappa_4(\lambda_3-\lambda_1)$ $+ \varkappa_3\varkappa_4(\lambda_1-\lambda_2)$
		$(1^2+; 1^2-)$	$\chi_1 = \varkappa_1\varkappa_2(\lambda_1-\lambda_2)(\lambda_3-\lambda_4) + \varkappa_2\varkappa_3(\lambda_2-\lambda_3) \times$ $\times (\lambda_1-\lambda_4) + \varkappa_3\varkappa_1(\lambda_3-\lambda_1)(\lambda_2-\lambda_4)$ $+ \varkappa_1\varkappa_4(\lambda_1-\lambda_4)(\lambda_3-\lambda_2) + \varkappa_2\varkappa_4(\lambda_2-\lambda_4) \times$ $\times (\lambda_1-\lambda_3) + \varkappa_3\varkappa_4(\lambda_3-\lambda_4)(_2\lambda-\lambda_1)$

Table 4 (continued)

Skeleton	Rep. of		$\mathscr{C}$-Chirality functions
	$\mathfrak{S}_n^i$	$\bar{\mathfrak{S}}_n$	
			$\chi_2 = \varkappa_1\varkappa_2[\Theta(2,4) - \Theta(1,4) - \Theta(2,3)$ $+ \Theta(1,3)]$ $+ \varkappa_2\varkappa_3[\Theta(3,4) - \Theta(2,4) - \Theta(3,1)$ $+ \Theta(2,1)]$ $+ \varkappa_3\varkappa_1[\Theta(1,4) - \Theta(3,4) - \Theta(1,2)$ $+ \Theta(3,2)]$ $+ \varkappa_1\varkappa_4[\Theta(4,2) - \Theta(1,2) - \Theta(4,3)$ $+ \Theta(1,3)]$ $+ \varkappa_2\varkappa_4[\Theta(4,3) - \Theta(2,3) - \Theta(4,1)$ $+ \Theta(2,1)]$ $+ \varkappa_3\varkappa_4[\Theta(4,1) - \Theta(3,1) - \Theta(4,2)$ $+ \Theta(3,2)]$ with $\Theta(j,k) = \Theta(k,j)$
		$(2,1^2-)$	$\chi_1 = \varkappa_1\varkappa_2\varkappa_3\varkappa_4(\lambda_1-\lambda_2)(\lambda_2-\lambda_3)(\lambda_3-\lambda_1)$ $\chi_2 = \varkappa_1\varkappa_2\varkappa_3\varkappa_4[\omega(1,2) + \omega(2,3) + \omega(3,1)]$ with $\omega(j,k) = -\omega(k,j)$
	$(1^4)_g$	(1^4+)	$\chi_1 = (\lambda_1-\lambda_2)(\lambda_1-\lambda_3)(\lambda_1-\lambda_4)(\lambda_2-\lambda_3)$ $(\lambda_2-\lambda_4)(\lambda_3-\lambda_4)$
(ii)	$(1^4)_g$	(1^4+)	$\chi_2 = \omega(1,2,3) - \omega(1,2,4) + \omega(1,3,4)$ $- \omega(2,3,4)$ with $\omega(j,k,l)$ totally antisymmetric
		$(1^2+; 1^2-)$	$\chi_1 = \varkappa_1\varkappa_2(\lambda_1-\lambda_2)(\lambda_3-\lambda_4) + \varkappa_2\varkappa_3(\lambda_2-\lambda_3)$ $(\lambda_1-\lambda_4) + \varkappa_3\varkappa_1(\lambda_3-\lambda_1)(\lambda_2-\lambda_4)$ $+ \varkappa_1\varkappa_4(\lambda_1-\lambda_4)(\lambda_2-\lambda_3) + \varkappa_2\varkappa_4(\lambda_2-\lambda_4)$ $(\lambda_3-\lambda_1) + \varkappa_3\varkappa_4(\lambda_3-\lambda_4)(\lambda_1-\lambda_2)$ $\chi_2 = \varkappa_1\varkappa_2[\Theta(2,4) - \Theta(1,4) - \Theta(2,3)$ $+ \Theta(1,3)]$ $+ \varkappa_2\varkappa_3[\Theta(3,4) - \Theta(2,4) - \Theta(3,1)$ $+ \Theta(2,1)]$ $+ \varkappa_3\varkappa_1[\Theta(1,4) - \Theta(3,4) - \Theta(1,2)$ $+ \Theta(3,2)]$ $- \varkappa_1\varkappa_4[\Theta(4,2) - \Theta(1,2) - \Theta(4,3)$ $+ \Theta(1,3)]$ $- \varkappa_2\varkappa_4[\Theta(4,3) - \Theta(2,3) - \Theta(4,1)$ $+ \Theta(2,1)]$

Table 4 (continued)

Skeleton	Rep. of $\mathfrak{S}_n^l$	$\bar{\mathfrak{S}}_n$	$\mathscr{C}$-Chirality functions
			$-\varkappa_3\varkappa_4[\Theta(4,1) - \Theta(3,1) - \Theta(4,2) + \Theta(2,3)]$ with $\Theta(j,k) = \Theta(k,j)$
		(1^4-)	$\chi_1 = \varkappa_1\varkappa_2\varkappa_3\varkappa_4(\lambda_1-\lambda_2)(\lambda_1-\lambda_3)(\lambda_1-\lambda_4) \times$ $\times (\lambda_2-\lambda_3)(\lambda_2-\lambda_4)(\lambda_3-\lambda_4)$ $\chi_2 = \varkappa_1\varkappa_2\varkappa_3\varkappa_4[\omega(1,2,3) - \omega(1,2,4) + \omega(1,3,4) - \omega(2,3,4)]$ with $\omega(j,k,l)$ totally antisymmetric
(iii)	$(2^2)_g$	(2^2+)	$\chi_1 = (\lambda_1-\lambda_2)(\lambda_3-\lambda_4)$ $\chi_2 = \Theta(1,3) - \Theta(1,4) - \Theta(2,3) + \Theta(2,4)$ with $\Theta(j,k) = \Theta(k,j)$
		$(1^2+; 1^2-)$	$\chi_1 = 2(\varkappa_1\varkappa_2 + \varkappa_3\varkappa_4)(\lambda_1-\lambda_2)(\lambda_3-\lambda_4)$ $- (\varkappa_1\varkappa_3 + \varkappa_2\varkappa_4)(\lambda_3-\lambda_1)(\lambda_2-\lambda_4)$ $- (\varkappa_1\varkappa_4 + \varkappa_2\varkappa_3)(\lambda_1-\lambda_4)(\lambda_2-\lambda_3)$ $\chi_2 = 2(\varkappa_1\varkappa_2 + \varkappa_3\varkappa_4)[\Theta(1,3) - \Theta(1,4) - \Theta(2,3) + \Theta(2,4)]$ $- (\varkappa_1\varkappa_3 + \varkappa_2\varkappa_4)[\Theta(3,2) - \Theta(3,4) - \Theta(1,4) + \Theta(1,4)]$ $- (\varkappa_1\varkappa_4 + \varkappa_2\varkappa_3)[\Theta(1,2) - \Theta(1,3) - \Theta(4,2) + \Theta(4,3)]$ with $\Theta(j,k) = \Theta(k,j)$
		$(2+; 2-)$	$\chi_1 = (\varkappa_1-\varkappa_2)(\varkappa_3-\varkappa_4)$
		(2^2-)	$\chi_1 = \varkappa_1\varkappa_2\varkappa_3\varkappa_4(\lambda_1-\lambda_2)(\lambda_3-\lambda_4)$ $\chi_2 = \varkappa_1\varkappa_2\varkappa_3\varkappa_4[\Theta(1,3) - \Theta(1,4) - \Theta(2,3) + \Theta(2,4)]$ with $\Theta(j,k) = \Theta(k,j)$
(iii)	$(2^2)_u$	$(2,1+; 1-)$	$\chi_1 = \varkappa_1(2\lambda_2-\lambda_3-\lambda_4) + \varkappa_2(2\lambda_1-\lambda_3-\lambda_4)$ $+ \varkappa_3(2\lambda_4-\lambda_1-\lambda_2) + \varkappa_4(2\lambda_3-\lambda_1-\lambda_2)$
		$(1+; 2,1-)$	$\chi_1 = \varkappa_1\varkappa_2\varkappa_3(2\lambda_3-\lambda_1-\lambda_2) + \varkappa_1\varkappa_2\varkappa_4 \times$ $\times (2\lambda_4-\lambda_1-\lambda_2)$ $+ \varkappa_1\varkappa_3\varkappa_4(2\lambda_1-\lambda_3-\lambda_4) + \varkappa_2\varkappa_3\varkappa_4 \times$ $\times (2\lambda_2-\lambda_3-\lambda_4)$
	$(1^4)_g$	(1^4+)	$\chi_1 = (\lambda_1-\lambda_2)(\lambda_1-\lambda_3)(\lambda_1-\lambda_4)(\lambda_2-\lambda_3) \times$ $\times (\lambda_2-\lambda_4)(\lambda_3-\lambda_4)$

Table 4 (continued)

Skeleton	Rep. of $\mathfrak{S}_n'$	$\bar{\mathfrak{S}}_n$	$\mathscr{C}$-Chirality function
			$\chi_2 = \omega(1,2,3) - \omega(1,2,4) + \omega(1,3,4)$ $\quad - (2,3,4)$ with $\omega(j,k,l)$ totally antisymmetric
		$(1^2+;1^2-)$	$\chi_1 = (\varkappa_1\varkappa_2 + \varkappa_3\varkappa_4)(\lambda_1-\lambda_2)(\lambda_3-\lambda_4)$ $\quad + (\varkappa_1\varkappa_3 + \varkappa_2\varkappa_4)(\lambda_3-\lambda_1)(\lambda_2-\lambda_4)$ $\quad + (\varkappa_1\varkappa_4 + \varkappa_2\varkappa_3)(\lambda_2-\lambda_3)(\lambda_1-\lambda_4)$ $\chi_2 = (\varkappa_1\varkappa_2 + \varkappa_3\varkappa_4)[\Theta(1,3) - \Theta(1,4)$ $\quad - \Theta(2,3) + \Theta(2,4)]$ $\quad + (\varkappa_1\varkappa_3 + \varkappa_2\varkappa_4)[\Theta(3,2) - \Theta(3,4)$ $\quad - \Theta(1,2) + \Theta(1,4)]$ $\quad + (\varkappa_1\varkappa_4 + \varkappa_2\varkappa_3)[\Theta(2,1) - \Theta(2,4)$ $\quad - \Theta(3,1) + \Theta(3,4)]$ with $\Theta(j,k) = \Theta(k,j)$
		(1^4-)	$\chi_1 = \varkappa_1\varkappa_2\varkappa_3\varkappa_4(\lambda_1-\lambda_2)(\lambda_1-\lambda_3)(\lambda_1-\lambda_4)$ $\quad \times (\lambda_2-\lambda_3)(\lambda_2-\lambda_4)(\lambda_3-\lambda_4)$ $\chi_2 = \varkappa_1\varkappa_2\varkappa_3\varkappa_4[\omega(1,2,3) - \omega(1,2,4)$ $\quad + \omega(1,3,4) - \omega(2,3,4)]$ with $\omega(j,k,l)$ totally antisymmetric
(iv)	$(2^2)_u$	$(2,1+;1-)$	$\chi_1 = \varkappa_1(2\lambda_3-\lambda_2-\lambda_4) + \varkappa_2(2\lambda_4-\lambda_1-\lambda_3)$ $\quad + \varkappa_3(2\lambda_1-\lambda_2-\lambda_4) + \varkappa_4(2\lambda_2-\lambda_1-\lambda_3)$
		$(1+;2,1-)$	$\chi_1 = \varkappa_1\varkappa_2\varkappa_3(2\lambda_2-\lambda_1-\lambda_3) + \varkappa_1\varkappa_2\varkappa_4$ $\quad \times (2\lambda_1-\lambda_2-\lambda_4)$ $\quad + \varkappa_1\varkappa_3\varkappa_4(2\lambda_4-\lambda_1-\lambda_3) + \varkappa_2\varkappa_3\varkappa_4 \times$ $\quad \times (2\lambda_3-\lambda_2-\lambda_4)$
	$(2,1^2)_g$	$(2,1^2+)$	$\chi_1 = \lambda_1\lambda_2(\lambda_1-\lambda_2) + \lambda_2\lambda_3(\lambda_2-\lambda_3)$ $\quad + \lambda_3\lambda_4(\lambda_3-\lambda_4) + \lambda_4\lambda_1(\lambda_4-\lambda_1)$ $\chi_2 = \omega(1,2) + \omega(2,3) + \omega(3,4) + \omega(4,1)$ with $\omega(j,k,) = -\omega(k,j,)$
		$(2+;1^2-)$	$\chi_1 = \varkappa_1\varkappa_2(\lambda_1-\lambda_2) + \varkappa_2\varkappa_3(\lambda_1-\lambda_3) + \varkappa_3\varkappa_4 \times$ $\quad \times (\lambda_3-\lambda_4) + \varkappa_4\varkappa_1(\lambda_4-\lambda_1)$
		$(1^2+;2-)$	$\chi_1 = \varkappa_1\varkappa_2(\lambda_3-\lambda_4) + \varkappa_2\varkappa_3(\lambda_4-\lambda_1)$ $\quad + \varkappa_3\varkappa_4(\lambda_1-\lambda_2) + \varkappa_4\varkappa_1(\lambda_2-\lambda_3)$
(iv)	$(2,1^2)_g$	$(1^2+;1^2-)$	$\chi_1 = (\varkappa_1\varkappa_3 - \varkappa_2\varkappa_4)(\lambda_3-\lambda_1)(\lambda_2-\lambda_4)$ $\chi_2 = (\varkappa_1\varkappa_3 - \varkappa_2\varkappa_4)[\Theta(3,2) - \Theta(3,4)$

Table 4 (continued)

Skeleton	Rep. of $\mathfrak{S}_n^l$	Rep. of $\bar{\mathfrak{S}}_n$	$\mathscr{C}$-Chirality function
			$- \Theta(1,2) + \Theta(1,4)]$ with $\Theta(j,k) = \Theta(k,j)$
		$(2,1^2-)$	$\chi_1 = \varkappa_1\varkappa_2\varkappa_3\varkappa_4[\lambda_1\lambda_2(\lambda_1-\lambda_2) + \lambda_2\lambda_3(\lambda_2-\lambda_3)$ $+ \lambda_3\lambda_4(\lambda_3-\lambda_4) + \lambda_4\lambda_1(\lambda_4-\lambda_1)]$ $\chi_2 = \varkappa_1\varkappa_2\varkappa_3\varkappa_4[\omega(1,2) + \omega(2,3) + \omega(3,4)$ $+ \omega(4,1)]$ with $\omega(j,k) = -\omega(k,j)$
(v)	$(1^4)_g$	(1^4+)	$\chi_1 = (\lambda_1-\lambda_2)(\lambda_1-\lambda_3)(\lambda_1-\lambda_4)(\lambda_2-\lambda_3)$ $(\lambda_2-\lambda_4)(\lambda_3-\lambda_4)$ $\chi_2 = \omega(1,2,3) - \omega(1,2,4) + \omega(1,3,4)$ $- \omega(2,3,4)$ with $\omega(j,k,l)$ totally antisymmetric
		$(1^2+; 1^2-)$	$\chi_1 = (\varkappa_1\varkappa_2 + \varkappa_3\varkappa_4)(\lambda_1-\lambda_3)(\lambda_3-\lambda_4)$ $+ (\varkappa_1\varkappa_3 + \varkappa_2\varkappa_4)(\lambda_3-\lambda_1)(\lambda_2-\lambda_4)$ $+ (\varkappa_1\varkappa_4 + \varkappa_2\varkappa_3)(\lambda_1-\lambda_4)(\lambda_2-\lambda_3)$ $\chi_2 = (\varkappa_1\varkappa_2 + \varkappa_3\varkappa_4)[\Theta(1,3) - \Theta(1,4)$ $- \Theta(2,3) + \Theta(2,4)]$ $+ (\varkappa_1\varkappa_3 + \varkappa_2\varkappa_4)[\Theta(3,2) - \Theta(3,4)$ $- \Theta(1,2) + \Theta(1,4)]$ $+ (\varkappa_1\varkappa_4 + \varkappa_2\varkappa_3)[\Theta(1,2) - \Theta(1,3)$ $- \Theta(4,2) + \Theta(4,3)]$ with $\Theta(j,k) = \Theta(k,j)$
		(1^4-)	$\chi_1 = \varkappa_1\varkappa_2\varkappa_3\varkappa_4(\lambda_1-\lambda_2)(\lambda_1-\lambda_3)(\lambda_1-\lambda_4) \times$ $\times (\lambda_2-\lambda_3)(\lambda_2-\lambda_4)(\lambda_3-\lambda_4)$ $\chi_2 = \varkappa_1\varkappa_2\varkappa_3\varkappa_4[\omega(1,2,3) - \omega(1,2,4)$ $+ \omega(1,3,4) - \omega(2,3,4)]$ with $\omega(j,k,l)$ totally antisymmetric

C. Chiral Representations for 6-site Skeletons

Table 5 lists the representations of $\mathfrak{S}_6^l$ which are chiral for one or more of the skeletons (vi), (vii), and (viii), which skeleton(s) they are chiral with respect to (with a number in parentheses indicating how many times, if more than once, the representation appears), and the representations of $\bar{\mathfrak{S}}_6$ that they induce.

Table 5. $\mathscr{C}$-Chiral representations of $\mathfrak{S}_6^l$ and $\overline{\mathfrak{S}}_6$ for various skeletons

$\mathfrak{S}_6^l$ Reps.	Skeletons	$\overline{\mathfrak{S}}_6$ Representations
$(4,2)_g$	viii	$(4,2+)$, $(2^2+;2-)$, $(4+;2-)$, $(3,1+;2-)$, $(3,1+;1^2-)$, $(1^2+;3,1-)$, $(2+;3,1-)$, $(2+;4-)$, $(2+;2^2-)$, $(4,2-)$.
$(4,2)_u$	vi, vii, viii(2)	$(3,2+;1-)$, $(4,1+;1-)$, $(2,1+;2,1-)$, $(2,1+;3-)$, $(3+;3-)$, $(3+;2,1-)$, $(1+;4,1-)$, $(1+;3,2-)$.
$(4,1^2)_g$	vi, vii, viii	$(4,1^2+)$, $(2,1^2+;2-)$, $(4+;1^2-)$, $(3,1+;2-)$, $(3,1+;1^2-)$, $(1^2+;3,1-)$, $(2+;3,1-)$, $(1^2+;4-)$, $(2+;2,1^2-)$, $(4,1^2-)$.
$(3,2,1)_g$	viii	$(3,2,1+)$, $(3,1+;1^2-)$, $(3,1+;2-)$, $(2^2+;2-)$, $(2^2+;1^2)$, $(2,1^2+;2-)$, $(2,1^2+;1^2-)$, $(1^2+;2,1^2-)$, $(2+;2,1^2-)$, $(1^2+;2^2-)$, $(2+;2^2-)$, $(2+;3,2-)$, $(1^2+;3,1-)$, $(3,2,1-)$.
$(3,2,1)_u$	viii	$(3,2+;1-)$, $(3,1^2+;1-)$, $(2^2,1+;1-)$, $(3+;2,1-)$, $(2,1+;2,1-)$, $(1^3+;2,1-)$, $(2,1+;1^3-)$, $(2,1+;3-)$, $(1+;2^2,1-)$, $(1+;3,1^2-)$, $(1+;3,2-)$.
$(3,1^3)_g$	vi, vii, viii(2)	$(3,1^3+)$, $(1^4+;2-)$, $(3,1+;1^2-)$, $(2,1^2+;2-)$, $(2,1^2+;1^2-)$, $(1^2+;2,1^2-)$, $(2+;2,1^2-)$, $(1^2+;3,1-)$, $(2+;1^4-)$, $(3,1^3-)$.
$(3,1^3)_u$	viii	$(2,1^3+;1-)$, $(3,1^2+;1-)$, $(1^3+;3-)$, $(1^3+;2,1-)$, $(2,1+;2,1-)$, $(2,1+;1^3-)$, $(3+;1^3-)$, $(1+;3,1^2-)$, $(1+;2,1^3-)$.
$(2^3)_g$	vii	(2^3+), $(2^2+;2-)$, $(2,1^2+;1^2-)$, $(1^2+;2,1^2-)$, $(2+;2^2-)$, (2^3-).
$(2^3)_u$	vi, vii, viii(2)	$(2^2,1+;1-)$, $(1^3+;1^3-)$, $(2,1+;2,1-)$, $(1+;2^2,1-)$.
$(2,1^4)_u$	vi, viii	$(1^5+;1-)$, $(2,1^3+;1-)$, $(2,1+;1^3-)$, $(1^3+;1^3-)$, $(1^3+;2,1-)$, $(1+;2,1^3-)$, $(1+;1^5-)$.

D. Properties of Chiral Representations

Table 6 lists, for the representations chiral with respect to skeleton (i), the order in λ and $\varkappa$ of the polynomial obtained by the first procedure of Section IV, and the number of ligands h in the function of the second procedure (alternate version). Table 7 does the same for the representations $\mathscr{C}$-chiral with respect to one or more of the skeletons (ii), (iii), (iv), and (v). As always, if the ligands are required to be achiral, only the representations without a (—) part in their diagram need be considered.

Table 6. $\mathscr{C}$-Chiral representations of $\overline{\mathfrak{S}}_3$ for C_{3v} symmetry

Representation	$g(\lambda)$	$g(\chi)$	h
(1^3+)	3	0	2
$(1+;1^2-)$	1	2	0

Table 7. $\mathscr{C}$-Chiral representations of $\overline{\mathfrak{S}}_4$ for various symmetries

Representation	Skeletons	$g(\lambda)$	$g(\chi)$	h
(2^2+)	iii	2	0	2
$(3+;1-)$	ii	0	1	0
$(2,1^2+)$	ii, iv	3	0	2
$(2,1+;1-)$	ii, iii, iv	1	1	1
(1^4+)	ii, iii, v	6	0	3
$(2+;2-)$	iii	0	2	0
$(1^2+;2-)$	ii, iv	1	2	1
$(2+;1^2-)$	ii, iv	1	2	0
$(1^2+;1^2-)$	ii(2), iii(2), iv, v	2	2	1
$(1+;3-)$	ii	0	3	0
$(1+;2,1-)$	ii, iii, iv	1	3	0
(2^2-)	iii	2	4	0
$(2,1^2-)$	ii, iv	3	4	0
(1^4-)	ii, iii, v	6	4	0

E. Properties of Skeletons

Table 8 lists, for the skeletons i—vii, the category ((a) or (b) of Section VIII) to which each belongs, as well as the chirality numbers of Section VII.

Table 8. Properties of skeletons

Skeleton	Category	o	u	o_{min}	u_{max}	o^+	$\bar{o}$	$\bar{u}$
i	a	1	3	1	3	1	2	2
ii	a	2	3	1	4	3	4	1
iii	a	2	2	1	4	2	4	4
iv	b	2	3	2	3	2	3	2
v	a	1	4	1	4	1	2	2
vi	b	4	3	3	4	4	6	1
vii	b	4	3	2	4	4	6	1
viii	b	4	3	3	4	4	6	1

XII. References

[1] Crum Brown, Proc. Roy. Soc. Edin. *17*, 181 (1890).
[2] Guye, Compt. Rend. *110*, 714 (1890).
[3] Boys, S. F.: Proc. Roy. Soc. A *144*, 655 (1934).
[4] Ruch, E., Schönhofer, A., Ugi, I.: Thoret. Chim. Acta *7*, 420 (1967).
[5] — — Theoret. Chim. Acta *10*, 91 (1968).
[6] — — Theoret. Chim. Acta *19*, 225 (1970).
[7] Mead, A., Ruch, E., Schönhofer, A.: Theoret. Chim. Acta *29*, 269 (1973).
[8] Ruch, E.: Accts. Chem. Research *5*, 49 (1972).
[9] Boerner, H.: *Darstellungen von Gruppen*, zweite Auflage. Berlin–Heidelberg–New York: Springer 1967. English translation: *Representations of Groups*, 2nd edition. Amsterdam: North-Holland Publishing Co. 1970.
[10] Young, A.: Proc. London Math. Soc. (2) *31*, 273 (1930).
[11] Hamermesh, M.: *Group Theory*. Reading, Mass.: Addison-Wesley 1962.
[12] Wigner, E.: *Gruppentheorie und ihre Anwendung auf die Quantenmechanik der Atomspektren*. Braunschweig: Friedr. Vieweg and Sohn 1931. English translation: *Group Theory and its Application to the Quantum Mechanics of Atomic Spectra*. New York: Academic Press 1959.
[13] Ruch, E., Schönhofer, A.: Theoret. Chim. Acta *3*, 291 (1965).
[14] Weyl, H.: *The Classical Groups*. Princeton: Princeton University Press 1946.
[15] Frame, J. S.: Nagoya Math. J. *27*, 585 (1966).
[16] Ruch, E.: Theoret. Chim. Acta *11*, 183 (1968).
[17] Haase, D., Ruch, E.: Theoret. Chim. Acta *29*, 189 (1973).
[18] — — Theoret. Chim. Acta *29*, 247 (1973).
[19] Richter, W. J., Richter, B., Ruch, E.: Angew. Chem. *85*, 21 (1973). English translation: Angew. Chem. Intern. Edit. *12*, 30 (1973).
[20] Ruch, E., Runge, W., Kresze, G.: Angew. Chem. *85*, 10 (1972). English translation: Angew. Chem. Internat. Edit. *12*, 20 (1973).
[21] Huppert, B.: *Endliche Gruppen I*. Berlin-Heidelberg-New York: Springer 1967.
[22] Kerber, A.: *Representation Theory of Permutation Groups I*. Berlin-Heidelberg-New York: Springer 1971.

Theoretica Chimica Acta

Editorial Board: C.J. Ballhausen, Copenhagen; R.D. Brown, Clayton; H. Hartmann (Editor-in-Chief), Mainz; E. Heilbronner, Basle; J. Jortner, Tel-Aviv; M. Kotani, Tokyo; J. Koutecký, Berlin; J.W. Linnett, Cambridge; E.E. Nikitin, Moscow; R.G. Pearson, Evanston; B. Pullman, Paris; K. Ruedenberg, Ames; C. Sandorfy, Montreal; M. Simonetta, Milan; O. Sinanoğlu, New Haven

Subscription Information and sample copies upon request

Springer-Verlag
Berlin
Heidelberg
New York

München Johannesburg
London Madrid
New Delhi Paris
Rio de Janeiro Sydney
Tokyo Utrecht Wien

Theoretica Chimica Acta publishes papers that are concerned with the relationship between individual chemical and physical phenomena and seeks links with the deductions of electron and valence theory. Papers on the theory of processes are invited. The contributions deal mainly with new research results, but they also report on experimental work in progress. In addition to original papers and short reviews, there are also occasional surveys.

Fields of Interest: Theoretical Chemistry, Atomic and Molecular Physics, Spectroscopy. Of Interest to: Chemistry Libraries and Institutes.

HMO
Hückel Molecular Orbitals

E. Heilbronner
and P.A. Straub

With 816 pages
DIN A 4
Loose Leaf. 1966
DM 92,—
US $35.50

Prices are subject
to change without
notice

From the reviews:

"In 1961, when Streitwieser wrote Molecular Orbital Theory for Organic Chemists, he drew attention to the very rapid recent growth of interest in this field — seventy papers in the forties, 600 in the fifties, and a corresponding increase in the sixties. These π-electron molecular orbitals are usually represented as linear combinations of atomic orbitals (LCAO) with certain other approximations as introduced by Hückel. The enormous use of these Hückel MOs has now led to no less than three fullscale publications of tables of the relevant coefficients. The present volume, prepared by Prof. Heilbronner and Dr. Straub, is the latest attempt to provide the coefficients in the MOs and certain other dependent quantities in such a form as to be helpful to chemists who have no desire to make these calculations for themselves." (Nature)

Springer-Verlag
Berlin Heidelberg New York

München Johannesburg London Madrid New Delhi
Paris Rio de Janeiro Sydney Tokyo Utrecht Wien